Eel Culture

Eel Culture

by Atsushi Usui

Translated by Ichiro Hayashi
Fish Hayashi Co. Ltd, Kanagawa Prefecture

SECOND EDITION

Fishing News Books

Fishing News Books
A division of Blackwell Scientific
Publications Ltd
Editorial offices:
Osney Mead, Oxford OX2 0EL
25 John Street, London WC1N 2BL
23 Ainslie Place, Edinburgh EH3 6AJ
3 Cambridge Center, Cambridge,
MA 02142, USA
54 University Street, Carlton,
Victoria 3053, Australia

First Edition published 1974
Reprinted 1979
Second Edition published 1991

Set by Setrite Typesetters Ltd, Hong Kong
Printed and bound in Great Britain by
The University Press, Cambridge.

DISTRIBUTORS
Marston Book Services Ltd
PO Box 87
Oxford OX2 0DT
(*Orders*: Tel: 0865 240201
Fax: 0865 721205
Telex: 83355 MEDBOK G)

USA
Blackwell Scientific Publications Inc
3 Cambridge Center
Cambridge, MA 02142
(*Orders*: Tel: (800) 759 6102

Canada
Oxford University Press
70 Wynford Drive
Don Mills
Ontario M3C 1J9
(*Orders*: Tel: (416) 441–2941)

Australia
Blackwell Scientific Publications
(Australia) Pty Ltd
54 University Street
Carlton, Victoria 3053
(*Orders*: Tel: (03) 347–0300)

British Library
Cataloguing in Publication Data
Usui, Atsushi
Eel culture.
1. Eels. Culture
I. Title
639.3751
ISBN 0–85238–182–4

Contents

List of Illustrations

List of Tables

Preface

Japan has cultivated eels since the late nineteenth century, and now produces roughly half the world's supply of eels. The open-pond systems which were traditionally used for intensive aquaculture have been superseded in the last twenty years by temperature-controlled glasshouses, in which eels can be raised to marketable size in half the time taken in open ponds. This has opened up the prospect of commercial eel production in countries where formerly it would not have been viable.

Atsushi Usui, a distinguished biologist and specialist in the techniques of eel culture, has updated his authoritative description of Japanese eel farming to take account of these developments. He explains the methods of eel production in simple terms, with many photographs and diagrams to illustrate the life cycle and habits of the eel, supply and demand worldwide, and the specialised methods of cooking eels in various countries.

This new edition provides aquaculturists with a reference work on the geographical and weather conditions for eel culture and its commercial possibilities worldwide.

The first edition of Eel Culture was translated by Ichiro Hayashi, and adapted for the world market by the British biologist Dr Gordon Williamson. This second edition has once again been translated by Mr Hayashi. The publishers would like to thank Professor R. H. Richards of the Institute of Aquaculture at the University of Stirling for his advice during the preparation of this new edition, and the Danish Carlsberg Foundation for their permission to reproduce Dr Johannes Schmidt's 1909 photograph of the stages of metamorphosis from leptocephalus to elver.

Conversion Factors

The conversion factors used in the book are:
1 hectare (ha) = 10 000 m^2 = 2.5 acres
1 m^3 = 220 gallons
1 metric ton (tonne, t) = 1000 kg = 2240 pounds

1 Introduction

Eel culture was started in Japan in 1894 and now the huge eel culture industry of that country produces about 37 000 t of eels per year and is still growing. Taiwan and Korea have copied Japan's methods and in Europe and in Australia and New Zealand there is much interest in getting eel culture started.

We must note that in a protein-hungry world it is highly wasteful to culture carnivorous animals such as eels, but that is the way it goes. An eel must eat 7 kg of fresh fish or 1.3–1.5 kg of the commercial diet in order to gain 1 kg body weight itself. This is usually composed mainly of white fish, which gives the best results. However, in Peru brown fish meal, made from anchovy, is used.

The Japanese eel species is called *Anguilla japonica*, or *unagi* in Japanese, and is one of 16 species of *Anguilla* in the world. All these eels have to a great extent similar habits and can undoubtedly be cultured by the same method as is used in Japan. The Japanese eel (*A. japonica*), European eel (*A. anguilla*), American eel (*A. rostrata*) and Australian and New Zealand eel (*A. australis*) are so alike that they are identical from a marketing point of view.

Table 1.1 shows the weights of eels caught and eaten in various areas of the world. About 18 000 t of wild eels and 99 000 t of cultured eels are harvested in the world in average years. The main eel-eating areas are Europe and the Far East (Table 1.2); in many other areas of the world where eels occur they are not eaten because people are afraid of their snake-like appearance. The main catches of eels are made in Europe, North America, the Far East, Australia and New Zealand.

The bodies of water in the world giving the heaviest sustained catches of wild eels are given in Table 1.3. Although many species of *Anguilla* occur in the tropics, actual stocks of eels and catches are only small in tropical areas.

The flesh of eels consists of delicious sweet meat with only one bone down the middle and there are at least three styles of cooking:

- smoked, the method used on the continent of Europe,
- jellied or stewed, the London method, and
- *kabayaki*, the Japanese style.

Almost every eel-producing area of the world is now exploited to yield the thousands of tonnes of eels eaten in Europe each year. Tanker-lorries bring live wild-caught eels from all over Europe, North

Table 1.1 World production of eels per year

Average of 1984–87 data from FAO and other sources. World annual production is about 18 000 t of wild eels and 99 000 t of cultured eels. Consumption rating: ***high, **medium, *low. The USA is the only area of the world where wild eel catches could still be significantly increased.

Area/Country		Wt caught per year (t)	Consumption rating	Net importer or exporter
Europe and N. Africa (*A. anguilla*)				Importer
Italy		3 800	*	Exported
France		2 500		Exported
Denmark		1 400	***	Imported
Poland		1 000	*	Balanced
Sweden		900	**	Balanced
N. Ireland and Eire		700		Exported
Netherlands		600	***	Imported
USSR		400	*	Balanced
Norway		300		Exported
Spain		300		Balanced
West Germany		150	***	Imported
Tunisia		120		Exported
Morocco		120		Exported
East Germany		100	**	Balanced
England, Wales, Scotland		80	*	Imported
North America (*A. rostrata*)				Exporter
Canada		700		Exporter
USA		400		Balanced
Asia (mainly *A. japonica*)				
Japan	wild	1 500	***	Importer
	cultured	37 000		
Taiwan	cultured	52 000	*	Exporter
S. Korea	cultured	3 000	*	Exporter
China	wild	3 000	**	Exporter
	cultured	7 000		
Australia (*A. australis*)		250		Exporter
New Zealand (*A. australis*) (*A. dieffenbachi*)		1 000		Exporter

Table 1.2 Main eel eating countries and their preferences

Ranking	Country	Preferred size & type of eel: Weight range (g)	Silver or brown	Style eaten	Approximate wholesale price £/kg 1989
1	Japan	150–200	cultured	*Kabayaki*	8.22 (cf. ¥1850/kg) 1 £ = ¥225 1989
2	W. Germany	over 260	Silver	Smoked	6.67
3	China	120–180	Brown or silver		
4	Netherlands	125–150	Silver	Smoked	7.50
5	France	50–260	Brown	Smoked	2.33
6	Denmark	300–400	Brown or silver	Smoked	6.27
7	Sweden	over 450	Silver	Smoked (sliced)	4.34
8	E. Germany	over 260	Silver	Smoked	
9	Britain	110–350	Silver or brown	Cold jellied or hot stewed	2.44*

This * price indicated in 1990.

cf.) Wholesale prices of eels in Japan in 1989 were as follows:

Jan.	¥1 750/kg	Jul.	¥2 100/kg
Feb.	¥1 750/kg	Aug.	¥1 900/kg
Mar.	¥1 800/kg	Sep.	¥1 750/kg
Apr.	¥1 900/kg	Oct.	¥1 600/kg
May.	¥2 150/kg	Nov.	¥1 650/kg
Jun.	¥2 200/kg	Dec.	¥1 700/kg

The prices of eels in Japan are not so low in Summer as in Europe because demand is increased in that season.

Table 1.3 Waters yielding the heaviest sustained catches of wild eels

Body of water	Eel catch per year Average (t)	Area km^2	Area Square miles
Ijssel Meer, Holland	2 300	1 800	700
Lough Neagh, N. Ireland	800	400	153
Commachio area, Italy	680	325	125
Lake Ellesmere, New Zealand, S. Island	300	260	100

Africa and North America while frozen wild eels are imported from New Zealand and Australia. Japan has not exported frozen cultured eels since 1968, because the local demand has increased with a corresponding increase in price.

The demand increases annually, as does the price! But there are no more unexploited wild stocks of eels in existence. Therefore the only way to obtain further supplies of this delicious food is to culture eels and the purpose of this book is to outline the practical methods that have evolved in Japan, in particular, for achieving that increased supply.

2 The world's species of *Anguilla*

The eels discussed in this book are members of a very unusual tropical genus of fish, *Anguilla*, which spend part of their lives in the sea and part in fresh water. There are thousands of eel species in the world that spend their whole lives in the sea and hundreds of species which live their whole lives in fresh water. But by a quirk of nature, only eels of the genus *Anguilla* have a life history half in the sea and half in fresh water.

The peculiar habits of *Anguilla* have mystified men for centuries. No one has ever found their ripe eggs. All other freshwater fish contain ripe eggs at some season, but never eels. How then do eels reproduce? Aristotle reckoned eels were created 'in the bowels of the earth'. English country folk believed elvers were fallen horse hairs that came to life when dropped in a puddle. Ranging from Europe to far Tahiti, magic origins were attributed to eels.

In fact, eels breed in the sea, far from land, and their sexual organs do not mature until after the adult eels migrate back from the rivers into the sea. That is why no eggs or sperm can ever be found in an eel captured in fresh water. Many ages ago *Anguilla* eels were marine fish like Conger eels, and later they somehow developed the habit of entering fresh water during the growing period of their lives.

The story of how the amazing life cycle of *Anguilla* was discovered is a major epic of science and to this day important aspects of the lives of eels still remain to be elucidated. The early scientific investigations on eels all centred on the European species *Anguilla anguilla*. This was due to the fact that in the nineteenth century scientific research existed only in Europe.

Two Italians, Grassi and Calandruccio, made the first important discovery in 1897 when they kept alive in a sea aquarium some peculiar transparent leaf-shaped fishes called *Leptocephalus* caught in the sea near Messina. To their amazement they watched these fishes during two months change their shape to become elvers of the fresh water eel (Figs 2.1 and 2.2). This showed that eels bred in the sea, and that the *Leptocephalus* was actually the larval stage of an eel. Grassi and Calandruccio suggested that the spawning place of eels was near Messina in the Mediterranean.

But they were wrong. The eel's amazing life migration remained a mystery for the Danish biologist Johannes Schmidt to elucidate.

As a young man then studying cod, Schmidt in 1904 was making plankton hauls to catch cod eggs far out at sea in the Atlantic off the Faroes when he found in his net a leptocephalus larva of an eel! Schmidt immediately realised that he had discovered a vital clue to

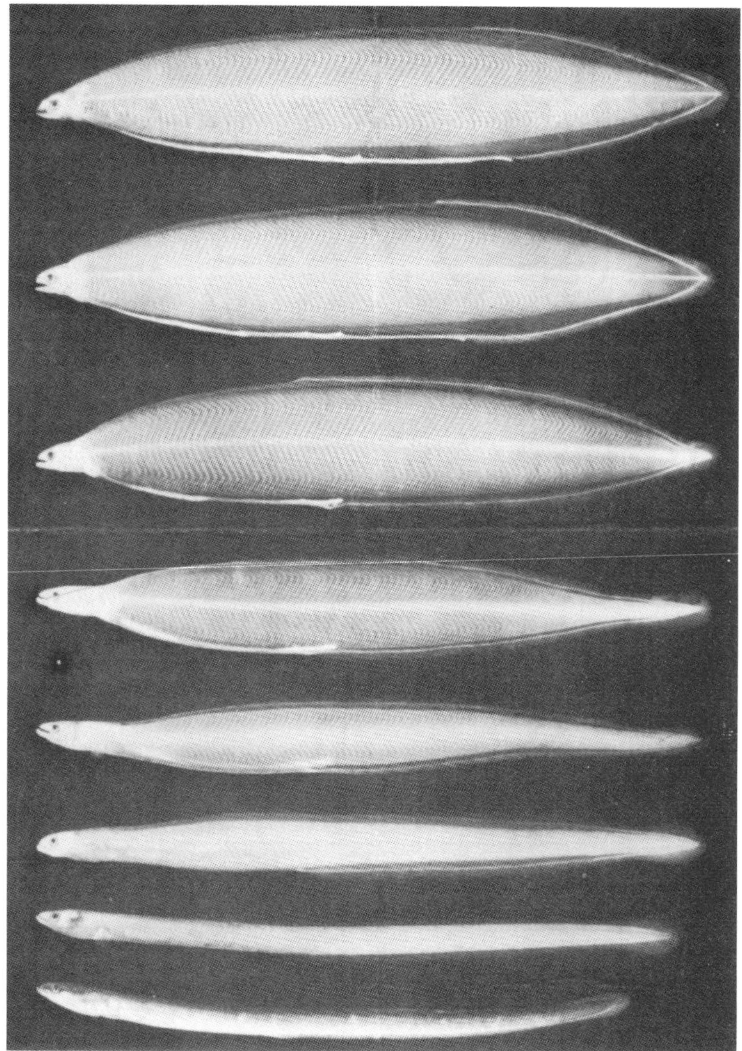

Fig. 2.1 Metamorphosis of a *leptocephalus* into an elver. This illustration, originally published in 1909 by the famous Johannes Schmidt, is historic. It established the development of the eel from larvae which, when alive, are completely transparent. The elver at the bottom represents the final transformation. Thanks are due to the Danish Carlsberg Foundation for permission to republish this historic illustration.

the life history of the eel. So, he thought, eel larvae occurred at Messina and off Faroe. Where else? That one larva started him on a research mission that lasted 35 years. With the backing of the Danish Government and the Carlsberg Foundation, Schmidt made cruises and plankton hauls all down the coasts of Europe. He caught many more eel larvae, all large ones. So, he wondered, in what part of the ocean did small larvae occur?

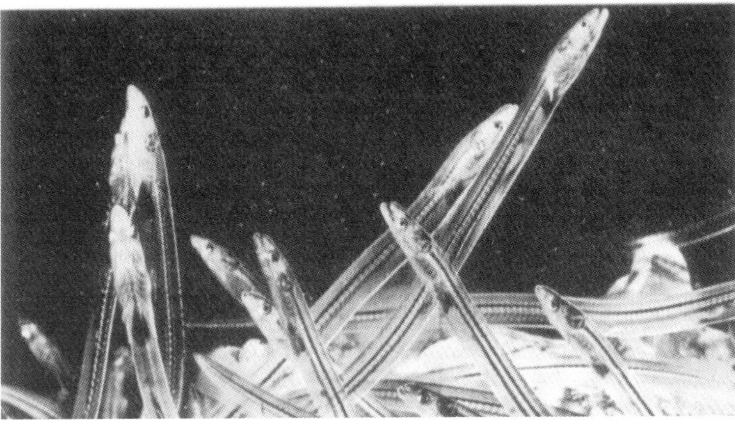

Fig. 2.2 Transparent elvers newly caught in the River Severn fresh from the sea. This is the starting point of eel culture. These elvers are already some 2¼ years old. Photo G. Williamson.

After World War I, Schmidt was able to make several cruises to the central and eastern parts of the Atlantic and in 1922 he announced the famous result of his persistent research. It was that the central part of the Atlantic called the Sargasso Sea was the only area where newly-hatched tiny eel larvae occurred (Fig. 2.3). The extraordinary European eel was thus shown to spawn in only one place, in the centre of the ocean far from land and in the tropical zone of the world (Schmidt, 1932). The ancestors of *Anguilla* were tropical marine eels which lived their whole life cycle in the sea.

Spawning takes place in February each year and is thought to occur about 400 metres beneath the surface, in water of about 17°C temperature (the eels spawn in mid water, not on the bottom). After being liberated by the adult eels the eggs rise and float near the surface, where they hatch in about 24 hours into tiny prelarvae 5 mm long. These tiny planktonic fish gradually grow into transparent, leaf-shaped leptocephalus larvae and get carried away from the Sargasso area by the Gulf Stream, which flows from that area away to the north east. After drifting for about 22 months the leptocephali arrive over the Continental shelf of Europe (which begins far from the actual coast) in November each year and there metamorphose into slim elvers. Possibly attracted by the smell of fresh water or some other driving force, these transparent elvers in millions head for the coast and enter river mouths. Because of temperature factors and the rate of current flow in different areas their time of arrival varies at different parts of the European coast. In the rivers of Northern Ireland, the southern and western rivers of England and Scotland and of Belgium, Denmark and Holland, the main arrival of elvers occurs in the months of April and May. Their size at the end of this journey is such that on average some 3500 are required to weigh 1 kg.

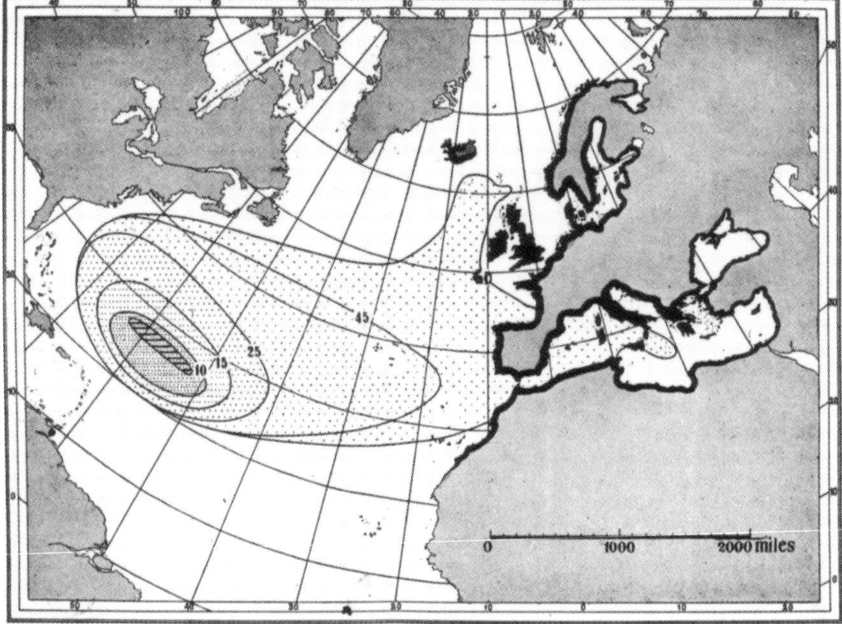

Fig. 2.3 Distribution of the European eel, *Anguilla anguilla*. The figure shows the end product of Johannes Schmidt's years of research which proved that eels spawned in the sea and were originally marine fishes. The only spawning ground for the European and the American eels is in the Sargasso area of the tropical Atlantic Ocean. The Gulf Stream carries leptocephalus larvae across to Europe and up the American coast. The figures on the map show their length in millimetres at each stage of their journey. The edge of the dotted zone marks the point at which leptocephali metamorphose into elvers. Note how leptocephalus penetrate the Mediterranean (after Schmidt, 1932).

Further south in the warmer waters of the French coast the arrival begins as early as February and the size is larger, needing only 2800 to the kg. Further south still on the coasts of Portugal and Spain the run begins as early as December and January and again, being fractionally larger because of having lost less weight they go 2700 to the kg.

The American eel (*A. rostrata*) also breeds in the Sargasso area and its elvers take about ten months to reach the American continental shelf.

In fresh water, the elvers turn black in colour and courageously migrate inland, nosing against the current at every stage, climbing around waterfalls by wriggling up the mossy sides of the falls (Fig. 2.4). They feed actively, mainly on insects and other small animals, and are wholly carnivorous; their diet is closely similar to that of trout. Although the great majority of eels spend their growing lives in fresh water, a minority stay in salt or brackish water coastal

Fig. 2.4 Elvers overcome waterfalls by wriggling up the wet mossy borders. They climb actively only at night; in the day they hide in crevices. This photo was taken in Victoria, Australia, and shows *Anguilla australis* elvers.

localities. Eels can be transferred from sea to fresh water and vice versa without any harm being caused.

When they are finally fully grown after various lengths of time, the eels migrate downstream into the sea in autumn and disappear from man's knowledge. Somehow, at least some of them must successfully navigate their way to the Sargasso spawning area. But no one has yet caught a single adult eel in the high seas. Nevertheless, it is deduced that they reach the Sargasso and spawn. Each female lays from 7 to 13 million eggs according to her size. Since no big eels ever re-enter rivers, we know that the adults must die after spawning. The ocean bottom of the Sargasso must be covered with their bones.

DIFFERENCES AMONG THE WORLD'S EEL SPECIES

His discovery made Schmidt famous. But he wanted further discoveries and so he immediately set his student Vilhelm Ege to try to discover how many other species of *Anguilla* existed in the world.

From museums he knew that many eels existed in the Indian and Pacific oceans. In fact, over-enthusiastic biologists had described over 100 species. By years of patient correspondence with museums, Danish consuls, missionaries, ships' captains and business men, persuading them to send collections of eels from every part of the world, and finally going themselves on a round-the-world eel-collecting expedition on the research ship *Dana*, Schmidt and Ege accumulated in Copenhagen a mammoth collection of 12 793 adult eels and 12 472 elvers from all over the world.

This vast collection was next examined in detail. Schmidt died in 1933 but Ege carried on. Finally, in 1939, Ege published the results that everyone had been waiting for. Truly both men gave their lives to eels.

Ege's list showed sixteen species of eels in the world and that the homeland of *Anguilla* in which they have evolved and from whence they have spread to other regions is the Indonesian Archipelago (Fig. 2.5).

The spawning places for South Pacific and Indian Ocean eels were investigated by Brunn (1937), Ege (1939) and Jespersen (1942). According to their studies the probable area for spawning was in the sea area ranging from the Bonin Islands to the Okinawa coast (far south of Japan), in depths of about 400 m, with temperatures above 13°C and salinity above 34‰. The collecting records of specimens of leptocephalus were as follows: 19 specimens at about 120 miles southeast of Okinawa Island in 1961 and one specimen at about 30 miles south of the southern edge of Taiwan in 1967.

Around the rest of the world, eels occur on most coasts which are washed by ocean currents that come from tropical latitudes. From such coasts adult eels migrate upstream to warm spawning grounds and the currents carry the larvae back to the coast. Off the western coasts of the southern continents, off North America and off the eastern coast of South America, ocean currents originate from cold polar seas or are otherwise unsuitable for larval drift, and so no eels occur on these coasts.

The names, characteristics and distribution of the world's species of *Anguilla* as listed by Ege are given in Table 2.1. The main characteristics that distinguish the species are:

- Colour mottled or plain, seven species are mottled, nine plain;
- Number of vertebrae range from 103 to 116;
- Dorsal fin length; 12 species have long fins, four species short fins. (Fig. 2.6);
- Maximum sizes to which eels grow range from 2.0 kg to 27 kg;
- Habitat preferred; still-water ponds or clear streams according to species;
- Distribution is shown in Table 2.1.

The range over which species are distributed varies greatly. The huge mottled *A. marmorata* (Fig. 2.7) occurs over a huge 12 000 mile curve of this planet's surface from South Africa to the Marquesas islands in mid-Pacific (Fig. 2.8). In contrast, *A. borneensis* is known only from eastern Borneo and Sulawesi and only five adults have been examined by scientists.

Anguilla japonica is the species of eel with the greatest number of vertebrae, 116 vertebrae on average. The common European eel *A. anguilla* has 115 vertebrae and the American eel *A. rostrata* averages 107 vertebrae. These three species of eels are very closely related and

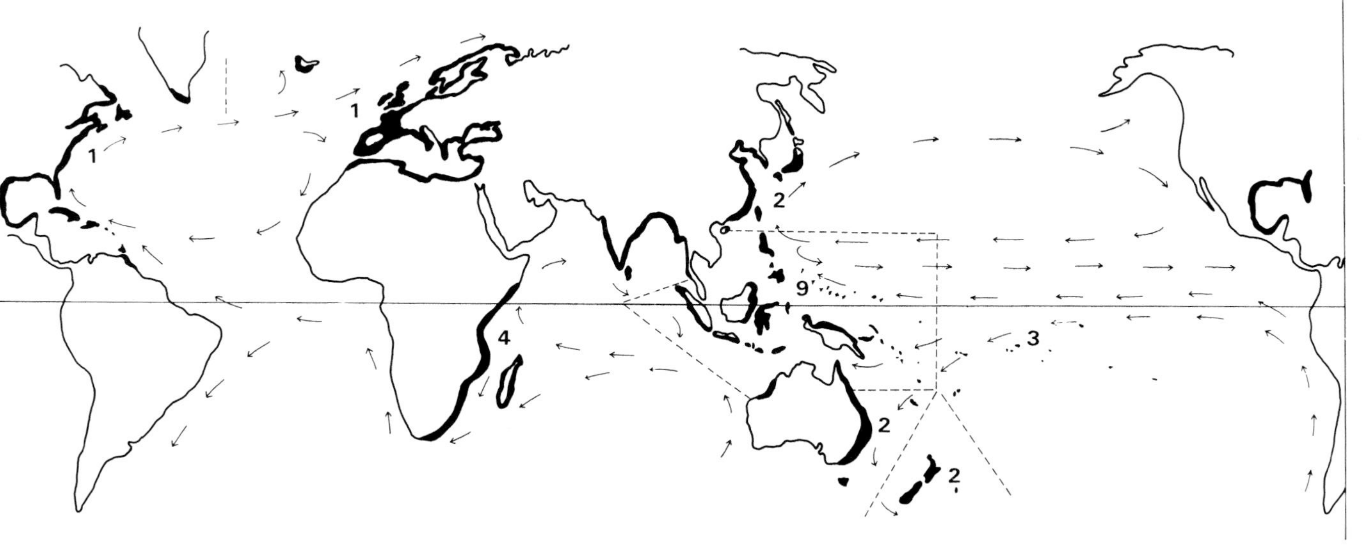

Fig. 2.5 The areas of the world inhabited by the various species of *Anguilla*. The areas concerned are in black. The major sea currents are indicated by arrows. The number of eel species found in various areas is marked; some species occur in several areas. The Indonesian Archipelago is the ancestral home of eels, and Ege (Schmidt's colleague) placed nine of the world's species as occurring in this area. Eels have never colonized the western coasts of the southern continents, western North America or eastern South America, because the ocean currents off these coasts do not provide any warm off-shore spawning area from which larvae can drift back to the coast. The cold currents at the bottom of Africa and South America are also unsuitable for eels.

Fig. 2.6 The dorsal fin of the long finned species of eels starts well in front of the anus. In the short finned species the dorsal and anal fin are more or less equal length.

undoubtedly have descended from a common ancestor. Apart from the number of vertebrae and the pattern of their teeth, there is little difference between these three species.

GROWTH OF BROWN AND SILVER STAGE EELS

During their growing life in fresh water, eels are called brown stage eels, yellow stage in American usage, on account of their brownish-yellow colour. Actually eels modify their colour to blend with their environment, and clear rivers have pale coloured eels, dark peaty ditches have dark eels and truly yellow eels occur only in certain habitats.

The abundance of wild eels in different bodies of water varies greatly. It is determined by the amount of food available, competition and predation by other fish and the numbers of elvers that reach that pond or stretch of river. Fast flowing rocky rivers have the lowest densities of eels, possibly only 5 kg per ha of river bottom. Slow flowing muddy rivers, alkaline lakes, fen-land drainage ditches and brackish water coastal lagoons have the highest density of eels. In certain brackish coastal lagoons in Newfoundland, 400 kg eels per ha (450 pounds eels per acre) are present.

Eels live from 2 to 30 years in fresh water before returning to the sea according to their species. Eels trapped in ponds or kept in aquariums have been known to live 85 years. The age of an eel can be determined by examining its otoliths (Fig. 2.9).

In the last summer of their life in fresh water, eels change to what is called the silver stage, preparatory to migrating back to the sea. Eels cease to feed as soon as they leave fresh water and thus must carry with them as oil a reserve of energy enough to power them on their long swim out to the ocean spawning grounds. Silver stage European eels, the species with the longest sea migration, contain no less than 28 per cent oil by weight. The greater oil content of silver stage than brown stage eels causes silver stage eels to fetch a high market price.

Table 2.1 Details of species of *Anguilla*, based on Ege (1939). Size data from various reports and field workers

Name	Colour of Body	Average number of vertebrae (counting hypural as being the last vertebra)	Distribution	Approximate maximum size of female eels (kg)	(cm)	Notes
A. ancestralis	Mottled	103	N. Sulawesi			
A. celebesensis	Mottled	103	Indonesia, Philippines			
A. interioris	Mottled	105	New Guinea			
A. megastoma	Mottled	112	Pacific islands from Solomons east to Pitcairn	22	190	
A. nebulosa	Mottled	110	East Africa and India	10	150	
A. marmorata	Mottled	106	South Africa, Madagascar, Indonesia, China, Japan, Pacific Islands	27	200	*A. marmorata* is the most widely distributed species
A. reinhardti	Mottled	108	Eastern Australia, New Caledonia	18	170	
A. borneensis	Plain	106	Borneo, Celebes	2	90	
A. japonica	Plain	116	Japan, China	6	125	*A. japonica*, *A. rostrata* and *A. anguilla* are closely related
A. rostrata	Plain	107	E. coasts of USA, Canada, Greenland	6	125	
A. anguilla	Plain	115	W. coast of Europe, N. Africa, Iceland	6	125	
A. dieffenbachi	Plain	113	New Zealand	20	150	
A. mossambica	Plain	103	South & East Africa, Madagascar	5	125	
A. bicolor	Plain	108	E. Africa, Madagascar, India, Indonesia, N. W. Australia	3	110	
A. obscura	Plain	104	New Guinea, Pacific islands from Solomons east to Tahiti			
A. australis	Plain	112	Eastern Australia & New Zealand	2.5	95	

Fig. 2.7 Mottled eel species, *A. marmorata*. Photo G. Williamson.

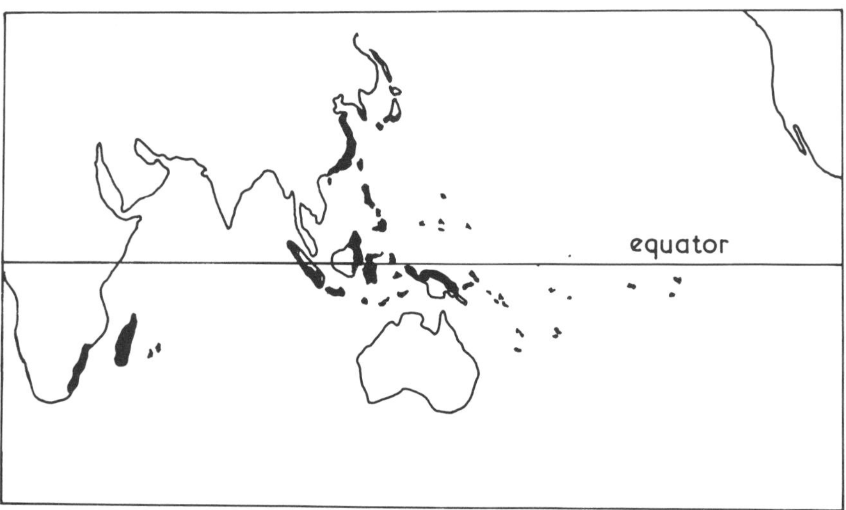

Fig. 2.8 The distribution of *A. marmorata* marked in black. It is the most widely distributed species, its range spanning 12 000 miles from South Africa to Japan and to the mid-Pacific Marquesas Islands.

The main characteristics of eels in their two stages are:

- **Brown stage (yellow stage)** body pigment dull, usually with a yellowy-green tinge, no sheen. Upper parts grey, brownish, greenish or yellowish. Underside dull white or grey. Fat content 5–15 per cent.

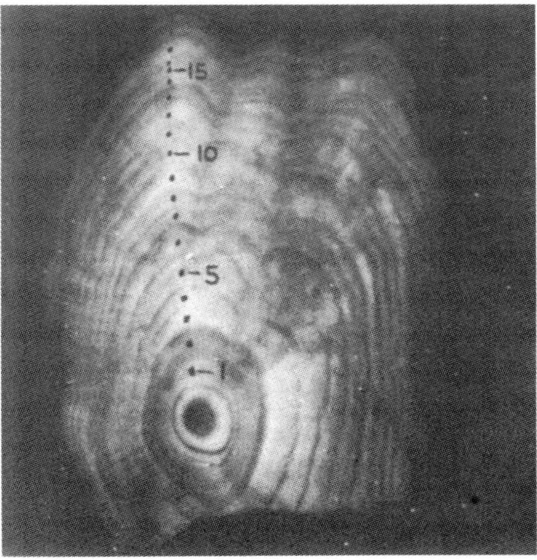

Fig. 2.9 The otolith of an eel reveals its age. Otoliths are calcareous lumps that grow in fishes' ears as part of the balancing apparatus. The rings reveal the growth of the fish like the annual rings in trees. This otolith came from an *Anguilla rostrata* caught in Newfoundland, Canada. The innermost opaque white zone was formed at sea during the leptocephalus stage. Subsequent zones are marked with dots and were formed in fresh water during each summer's growth. This eel had spent one year in the sea as a larva followed by 19 in fresh water, so it was 20 years old.

- **Silver stage**: A glistening layer exists underneath the skin. This causes the whole body to have a silvery sheen or glint, especially the white underside. Upper parts of the body are usually grey, occasionally with a purple sheen. No green or yellow tints remain.

In this stage the fat content is greater than in the brown stage and in the European eels reaches 25–28 per cent.

On autumn nights, especially during the last quarter of the moon, and when heavy rains have caused rivers to flood, the silver eels migrate downstream to the sea. By setting a net across a river, the entire run of migrating eels can be caught, and since adult eels never return from the ocean spawning grounds, the capture of these eels is totally harmless from a conservation point of view. Many other eels from lesser rivers that are not netted will always reach the spawning grounds and the elvers they produce are distributed by the ocean currents to the whole coastline (Figs 2.10, 2.11, 2.12).

FEMALE EELS ARE BIGGER THAN MALES

Female eels grow to much bigger sizes than do males and live longer in fresh water before returning to the sea. Since male eels are so small

Fig. 2.10 The use of fyke nets is one popular way of catching brown stage eels. Photo G. Williamson.

Fig. 2.11 Using baited eel pots is another successful method of catching brown stage eels. Photo G. Williamson.

Fig. 2.12 Illustrated here is one way of catching silver stage eels at night on the River Bann, N. Ireland. Such eels are most conveniently caught on migration into the sea in autumn. A permanent barrier of mesh or wattle can be constructed across the whole width of a river, leaving openings in which nets can be set. The eels travel only at night and men must empty the net frequently. On one exceptional night in 1963 on the River Bann 8 t of eels were caught. In the River Po area in North Italy permanent weirs and installations provide the basis for a large eel processing industry which has been famous for centuries. Photo G. Williamson.

they have no commercial significance and 90 per cent of the weight of wild eel catches anywhere in the world is made up of females. The average sizes and ages of migrating silver eels of various species are given in Table 2.2.

FANTASTIC 6500 KILOMETRE MIGRATION

How European silver eels from many distant parts of Europe such as the Baltic, France and even from Egypt in Africa all migrate back to the single Sargasso spawning area is a mystery. They seem to swim near the surface, for the crews of lightships at sea sometimes see eels swimming past. The silver bellies of the eels are a camouflage protection against predators only when seen from below against the sky.

Johannes Schmidt noted that all eel species spawn in the highest salinity zones of the oceans. And eels are able to detect very small changes in salinity, light and gravity, especially the gravitational pull of the moon.

After leaving river mouths and while still over shallow water, eels probably navigate using salinity, light and the gravitational forces of

Table 2.2 Size, age and period of migration of silver stage eels of various species

The figures given are the average values. Note that male silver eels return to the sea at much smaller sizes and about a month earlier than female silvers. Each species of eel returns to the sea when it reaches a certain size: the number of years taken to reach this size varies according to the growth rate in each locality which in turn is determined chiefly by the water temperature and amount of food available.

		Femàle silver eels				Male silver eels			
Species	Locality	wt (g)	length (cm)	age (years)	season of migration	wt (g)	length (cm)	age (years)	season of migration
A. anguilla	N. Ireland	260	50	12	Sep.–Oct.	120	40	9	Aug.
A. rostrata	Canada, St Lawrence River	1 600	91		Sep.–Oct.				
A. rostrata	Newfoundland, Topsail Pond	600	70	12	Sep.				Jul.–Aug.
A. japonica	Japan, Southern Honshu	200–250	58	7	Sep.–Oct	100–150	35	5	August
A. japonica	China, West River	270	55	3	Oct.–Nov.	140	46	3	Sep.–Oct.
A. marmorata	China, West River	12 000	150	12	Oct.–Jan.	1500	80	8	Sep.–Nov.
A. dieffenbachi	New Zealand, Lake Ellesmere	6 000	120	30	May	600	64	20	Apr.
A. australis	New Zealand, Lake Ellesmere	600	68	22	Apr.	200	46	14	Mar.
A. australis	Australia, Victoria, Lake Yambuk	600	70		Feb.–Mar.	200	45		Jan.–Feb.
A. reinhardti	Australia, Victoria	5 000	110		Mar.–Apr.				Feb.–Mar.

moon and sun to guide them. Once beyond the continental shelf and into the ocean current system they probably just head into water of ever increasing salinity. This alone will automatically cause them to arrive at the spawning ground. Many must die on the way.

How many years does the journey take from Europe up the Gulf Stream to the Sargasso? The distance is about 6500 km (4000 miles) and a 7 km/day current is against the eels. Experiments on tagging silver eels in the Baltic show that they swim about 20 km/day. They could thus average 13 km/day against the Gulf Stream and it would take them 1 year 5 months to swim 6500 km. This time seems correct since it means that silver eels leaving European rivers in autumn would reach the Sargasso in the second February following, which is the month when spawning occurs. During this long journey they do not feed. They must be living skeletons when they arrive, and be hardly recognisable as eels. In most other species of eel the spawning areas are much nearer the coast, and so the migration back to the spawning area must be much quicker.

Although no one has yet succeeded in catching eels at the ocean spawning grounds, it has been possible, by injecting hormones into female eels to induce their ovaries to develop almost up to full maturity (Fig. 2.13). Many eggs can be seen in the enlarged ovaries of such eels and this has enabled estimates of fecundity to be made.

HOW EELS FEEL AND BREATHE

Along the flanks of an eel and on its head are rows of pores. This system of pores is called the lateral line system and each pore contains cells that are sensitive to vibrations. An eel has two tubular nostrils which open into large sensory pits. Elvers swimming close to a river bank are easily frightened away by heavy footfalls, for example. Since an eel moves and feeds mainly in the half light of dawn and dusk, and often in murky water, it cannot see much with its eyes. Instead, the sensory cells of its nostrils and lateral line pores guide it, by enabling detection of smells and vibrations caused by any moving object.

Eels are smooth skinned and very slimy. Most people would say that an eel has no scales, but actually many very small scales are present embedded under the skin.

Owing to the absence of large scales, an eel can breathe through its skin as well as through its gills. The proportion of respiration (breathing) carried out through the gills is about 40 per cent and that through the skin about 60 per cent. On rainy autumn nights when silver stage eels feel the urge to migrate to the sea, they can easily wriggle through sodden fields to escape from land-locked ponds into nearby rivers; as long as their skin is wet they can get enough oxygen to live.

The oxygen consumption of eels increases as the temperature in-

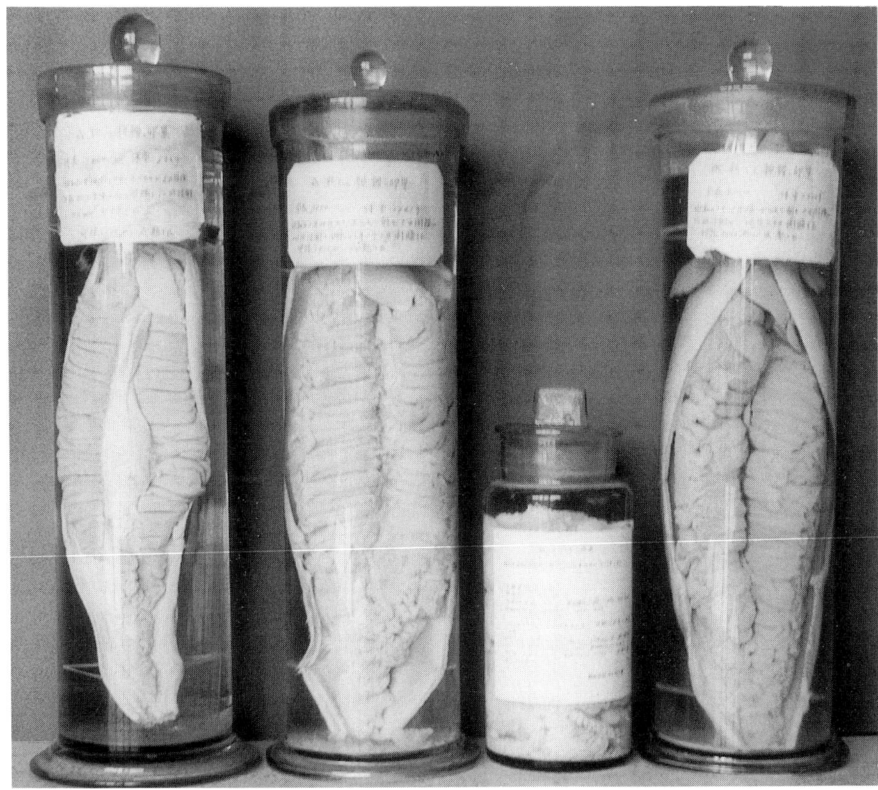

Fig. 2.13 Enlarged ovaries of female eels that have been injected with hormones. So far, no one has succeeded in inducing captive eels to breed. Hence eel culture depends on the capture of enough wild elvers each year.

creases, as is the case in all living things. This means that in winter a ton of eels have little requirement for oxygen and can be kept in a small pond. But in the heat of summer the same eels will need up to ten times the amount of oxygen.

MANY EEL MYSTERIES REMAIN

All the species of *Anguilla* undoubtedly have a similar life history, but the tropical species probably spawn quite near the coast and do not go to remote single spawning areas. The two Atlantic species are the odd members of the family, living far cut off from the ancestral zone, shivering out their lives in cold waters, yet showing their true preference for tropical seas by making the most fantastic migration known in biology to reach the warm tropical waters of the Sargasso to spawn. How eels ever got into the Atlantic is itself a mystery. Probably *A. anguilla* and *A. rostrata* originated as wandering members of the

Japanese eel species which somehow crossed from the Pacific into the Atlantic, perhaps through the sea channel which existed in the Panama region long ago.

Despite the solid foundation to ecological study of eels provided by Schmidt and Ege's work, there has been little progress in recent decades. The stage of knowledge at the present time is that the fresh water part of the life history, but not the marine part, is known in detail for the three northern eel species, *A. anguilla* and *A. rostrata* in the Atlantic and their close relative *A. japonica* in the N.W. Pacific. Dr Peter Castle of Victoria University, Wellington, has recently led an extensive investigation of the two New Zealand species. Almost nothing has been discovered about the tropical eel species, and wild eels with eggs have still never been seen.

About the marine part of eels' lives hardly anything is known. Where are the spawning grounds of each species and at what depth? By what remarkable method do adult eels navigate from many different points to reach the spawning area – surely a topic with many useful repercussions to man. By what route and at what depth do they swim?

Do tropical eel species spawn all year round or have they a single spawning season? Take the case of *Anguilla marmorata*, the king of eels, growing to 27 kg in size, whose realm extends from South Africa to Japan or to mid-pacific Tahiti and the Marquesas? (Figs 2.7 and 2.8). When and how does it spawn? Surely it must have at least three completely separate spawning areas?

Here are some projects that scientists might dream of tackling:

- To catch spawning adult eels of any species.
  Probably it would be easiest to try one of the three northern species first, since we have a good idea as to the spawning place and season of these three. Delicate sonar, underwater lights, electric paralysing methods, underwater TV or submarines, depth charge explosions and huge nets may be needed – and money! Probably the first man to see such eels will not recognise them as eels, they will be so emaciated.
- To discover the route and depth by which any adult eel travels from the coast to the spawning ground.
  Here there should be studied a big species, to which could be attached radio transmitter tags; one that spawns near to the coast, so that the tracking ship would not have to be chartered too long at too great expense. A big eel (*A. marmorata*) of S. Africa or Java would be a good choice.
- To search the bottom of the Sargasso Sea to see if it is covered with countless bones of the eels that are supposed to die after spawning there, grabs and dredge nets would be needed.
- Why do eels breed in the sea and enter rivers to feed whereas salmon breed in rivers and go to feed in the sea? Is it connected with the fact that the density of animal life in the sea is greater in

cold northern waters than in the tropics? Eels, being primarily tropical fish, may have discovered that there was little food available in the tropical sea, but plenty of food if they entered rivers. And perhaps salmon found that highland rivers yielded a meagre diet but that the sea beyond the river mouth was teeming with food.

There are many, many mysteries about eels that have yet to be solved.

3 The market demand for eels in Europe

About 25 000 t of eels are eaten each year in Europe. Eels must be marketed alive or quick-frozen and glazed in order to secure top prices. Generally speaking, eels caught in Europe are sold alive, eels imported from the USA and Canada are alive and eels imported from Australia and New Zealand are frozen. Unfrozen dead eels find almost no market. The price of eels is high and is rising steadily.

Who are the eel connoisseurs of Europe? In order of national consumption they are the Germans, Dutch, French Danes and Swedes. In all these countries smoked eels are on sale in every fish shop and are eaten in every household, ranking as an expensive treat similar in status to sole or turbot in Britain. Actually the Dutch and Danes eat the most eels per person but the much larger population of Germany makes it the nation that eats the greatest quantity of eels. In Belgium, Italy (at Christmas) and in London, Britain, lesser amounts of eels are eaten. The eel consumption in Britain is at present less than about 1000 t per year and is almost completely restricted to the East End of London. There is, however, a trend towards luxury eating in high-class clubs and restaurants, with the price equalling that of smoked salmon.

In Europe all eels over 50 g will find a sale but each country has its own size and type preferences (Table 1.2). In eating habits the Dutch are 'suckers' but the Germans are 'biters'. When a Dutchman eats an eel he likes to feel the oil trickling out of the corners of his mouth and down his chin. When a German eats an eel he likes to bite something big and solid. In this way, Joh Kuijten the founder of the famous eel firm Joh Kuijten of Spaarndam, Holland, explained to his sons the preferences of different countries. Holland and Germany are neighbours, yet, strange but true, they prefer quite different types of eels for smoking.

Which are the main centres of eel buying and selling in Europe? In order of importance, Spaarndam (headquarters of the giant eel business of Kuijten, which buys eels in Europe, Africa, the United States and most of the world for the European market), and Hamburg.

What factors affect the price of eels? The highest prices are obtained during the winter season December–April. This is when wild eels cannot be caught and so supplies are low, the lowest prices occur during the main catching season, June–October. Silver stage eels always fetch a better price than brown-stage eels of the same size, on account of their higher oil content and better flesh quality.

4 Life history of the Japanese eel *Anguilla japonica*

The Japanese eel *Anguilla japonica* spawns in the ocean near Okinawa. The larvae are carried toward the coasts by the sea currents and enter rivers as elvers; they ascend into fresh water rivers and lakes and live and feed there for 5–10 years until they reach adult size. After that they swim downstream into the sea and out to the spawning ground, where they spawn and die. As in all species of *Anguilla*, adult females are much heavier than adult males.

The land distribution of *Anguilla japonica* extends from northern Honshu through southern Japan, Korea, Taiwan and down the coast of China as far south as Hainan.

The part of the life cycle that occurs in fresh water is well-known. Each year during December–April, millions of elvers enter river mouths from the sea. The elvers are about 6 cm long and transparent and hence they are sometimes called 'glass eels'. They prefer to enter rivers when the river water temperature is 8–10°C, so the peak month of the immigration period varies somewhat between warm and cold years. The elvers have a strong instinct to swim against any water current and this causes them to swim vigorously upstream. When they encounter a waterfall they wriggle up the side of the falls among the wet mosses. They swim actively at night but in daytime hide under banks and stones, etc. After some two weeks in fresh water, elvers become black in colour and look like small eels for the first time.

By midsummer, the young eels have reached rivers, ponds and streams all over the countryside, and in the warm summer temperatures of 25–32°C feed actively on insects and worms and grow to lengths of around 15 cm.

This feeding and growth continues, in wild eels, for 5–10 years. Cultured eels grow much faster. Brown stage eels is the name given to eels during this fresh water growing phase of their lives owing to the fact that during this period their colour is brownish-yellow.

When the eel is over 30 cm in length, its sexual organs can be distinguished for the first time. Finally, the eels reach adult size; males average 70 g weight and 35 cm long, and females are 300–350 g weight and 57–60 cm long. Female eels are thus about four times as heavy as males, and the ratios of males to females in numbers are 4:6 in wild eels and 8 or 9:1 in cultured eels. (See Chapter 10 for reference to the feeding of hormones to eels to increase the ratio of females to males.)

In the summer of the year in which an eel reaches adult size, nature prepares it for its journey down to the sea and out to the distant ocean spawning grounds. Great amounts of oil are stored in the muscles, rising to about 20 per cent of the total body weight.

A metallic silvery colour develops under the skin of the eel, giving the belly a silvery-white appearance; the eels stop feeding, and on dark autumn nights the eels migrate down the river into the sea. Eels in this condition are called silver stage eels.

5 Still water and running water culture methods

The principles of eel culture are the same as for all fish culture. Water, elvers, feed, disease and marketing are the five main aspects:

- Water: for the eels to live in, a pure, plentiful supply is needed. Oxygen supply must be good and capable of being increased above the natural level for many eels to be cultured in a small area.
- Elvers: the harvest of eels is nothing more than elvers grown to market size.
- Feed: to grow the eels. The growth rate is determined by the temperature of the water and the amount of food eaten.
- Diseases and parasites: must be prevented and/or cured.
- Marketing: a man who knows how to market his eels will get twice the price for his fish compared with one who has no marketing experience.

Eels are cultured in ponds and live at much higher densities than can be supported by the oxygen supply available in an ordinary lake. To provide the extra oxygen that the eels need two fundamentally different methods exist and these give rise to two basic types of pond and culture techniques, the still water and the running water methods.

All Japanese eel ponds are of the still-water type and this is the method that can most easily be set up in most parts of the world. The running water method, as used at trout farms, is not used at present for eels but may be used in the future.

Before going further we must clearly describe and appreciate the characteristics of the two methods.

- Still water ponds. The first method of providing oxygen to the eels is by encouraging green phytoplanktonic algae to grow in the pond water which, by their photosynthetic activities, produce oxygen. Dense phytoplankton can develop only if the through-flow of water in the ponds is nil or very slow (if fast, the phytoplankton will simply be washed away). Thus ponds with more or less static water are required by this method. The daily through-flow of water in typical still water ponds is about 5 per cent of pond volume per day.
- Running water ponds. The second method of providing oxygen to the eels is to let continuous new supplies of oxygenated water enter the ponds, as at a trout farm. Thus these ponds must be supplied with plentiful running water all the time.

At any particular location, it is the volume and temperature of the water available that chiefly determines which type of pond is best; where only limited water is available, still water or recirculating running water ponds; where an unlimited supply of water is available, still water or, in tropics only, running water ponds.

TEMPERATURES OF 23–30°C NEEDED FOR COMMERCIAL SUCCESS

For commercial success, eels must reach market size of 150–200 g in two years or less. Really fast growth is needed to achieve this growth rate and eels grow fast only at temperatures of 23–30°C. The time taken for eels to reach market size is one and a half years in Taiwan but 4 years in England (see Table 14.4, p. 114). Below approximately 12°C *Anguilla japonica, A. anguilla* and *A. rostrata* do not feed and thus do not grow at all.

Note that eels require much warmer temperatures for growth than trout. The water temperature suitable for eel culture is 23–30°C, and for trout 10–15°C.

Thus outdoor unheated ponds can be used for commercial eel culture only in tropical and sub-tropical areas such as Indonesia, Taiwan, southern Japan, Madagascar, the Caribbean, Queensland and Tunisia.

In cooler areas such as northern Europe, southern Australia and New Zealand, eel ponds must be heated artificially if eels are to be grown commercially. There are several ways of doing this. Still water ponds may be covered with greenhouse polyethylene or glass, or alternatively pipes may be laid on the pond bottom through which hot water is circulated, i.e. pond central heating. Running water ponds may be covered and have insulated bottoms and walls, and the water is heated, purified and recirculated. This will be very expensive but is being attempted experimentally.

6 Methods and organization of the eel culture industry in Japan

Two species of eel are cultured in Japan, the local species *Anguilla japonica* (97 per cent of the cultured harvest in 1989) and the European species *A. anguilla* imported as elvers by air freight (three per cent of cultured harvest in 1989). The import of European elvers to Japan started about 1969 and they grow equally well as the Japanese species except that they do not like the 30°C temperatures which occur during the height of Japanese summer.

The eel ponds of Japan are not spread evenly around the country but are nearly all concentrated in three areas (Figs 6.1 and 6.2). The first is around the shores of the brackish lake Hamanako near Hamamatsu in Shizuoka Prefecture (*Hamana* = name, ko = lake, *Hamanako* = Lake *Hamana*). Another is Yoshida district by the Ohi river in Shizuoka Prefecture, about 100 miles south-west of Tokyo, extending to Yaizu district. The third is in Nishimikawa District, Aichi Prefecture (Fig. 6.3). All these areas are beside the coast.

All the eel ponds in Japan obtain their water supply from shallow boreholes and use the still water pond system (Fig. 6.4). The reason the still water system is used is as follows. Borehole water is 15–20°C all the year round, which is cold for eel culture. However, in still water ponds the sun heats the water up to 25–30°C, which is ideal for eel culture. This still water enables phytoplankton to propagate, giving suitable water darkness, and increases oxygen in the water, thus providing suitable conditions for eel culture.

In certain isolated eel farms in Japan, and in most eel farms in Taiwan, river water is used to irrigate the ponds. These farms also use the still water system. Young elvers, however, are reared in running-water tanks.

Eels are raised from elvers to market size using four sizes of ponds (Table 6.1). As the eels grow they are known by different names, which are described in Figs 6.5 and 6.6. The methods used to feed the eels of each size are described in Chapter 10. Fig. 6.6 shows the weights of eels of each different length. The maximum density at which eels can live in the most modern ponds is 4 kg/m^2 (Table 6.2).

Elvers at the transparent glass-eel stage are the starting point of eel culture. As yet, no one has succeeded in breeding eels and, as the natural spawning ground is far out to sea, it is necessary for elvers to be collected when they enter river mouths. They are about 6 cm long and transparent and weigh about 0.16 g each (it may take up to 6000 elvers to weigh 1 kg). On arrival at the farm they are put for six

Table 6.1 Details of ponds used to culture eels in Japan

Size of eels grown in pond	Approximate area of pond	Depth	Bottom	Covered or open	Temperature	Type of water supply	Notes	Approximate period in use
New elvers	20 m^2	0.6 m	concrete	under greenhouse	heated to above 25°C	Running water	Aerated by bubblers	Dec.–Mar.
Elvers 8–12 cm	30–100 m^2	1 m	concrete	under greenhouse	heated to above 25°C	Running water	Aerated by bubblers	Mar.–May
Fingerlings 12–20 cm	200–300 m^2	1 m	mud	greenhouse mainly, and out of doors in summer	heated to above 25°C and natural in summer	Still water	Overhanging lip to pond. Green water	Jun.–Jul.
Adults 20–70 cm	500–1000 m^2	1 m	mud	greenhouse mainly, and out of doors in summer	heated to above 25°C and natural in summer	Still water	Green water	mostly harvested in Aug. by use of greenhouse; remainder in summer of following year

Table 6.2 Weight of eels per m^2 of bottom that can live in still water, mud bottom ponds

These values apply to still water mud bottom ponds in which the eels burrow in winter to hibernate. Future running water, concrete bottom, artificially-heated ponds in which eels grow continuously may be stockable at higher densities. In this case, maximum oxygen must be supplied by using paddle splasher machine etc.

Pond size (m^2)	kg eels per m^2 sustainable
10 000	1.0
500	2.5
200	4.0

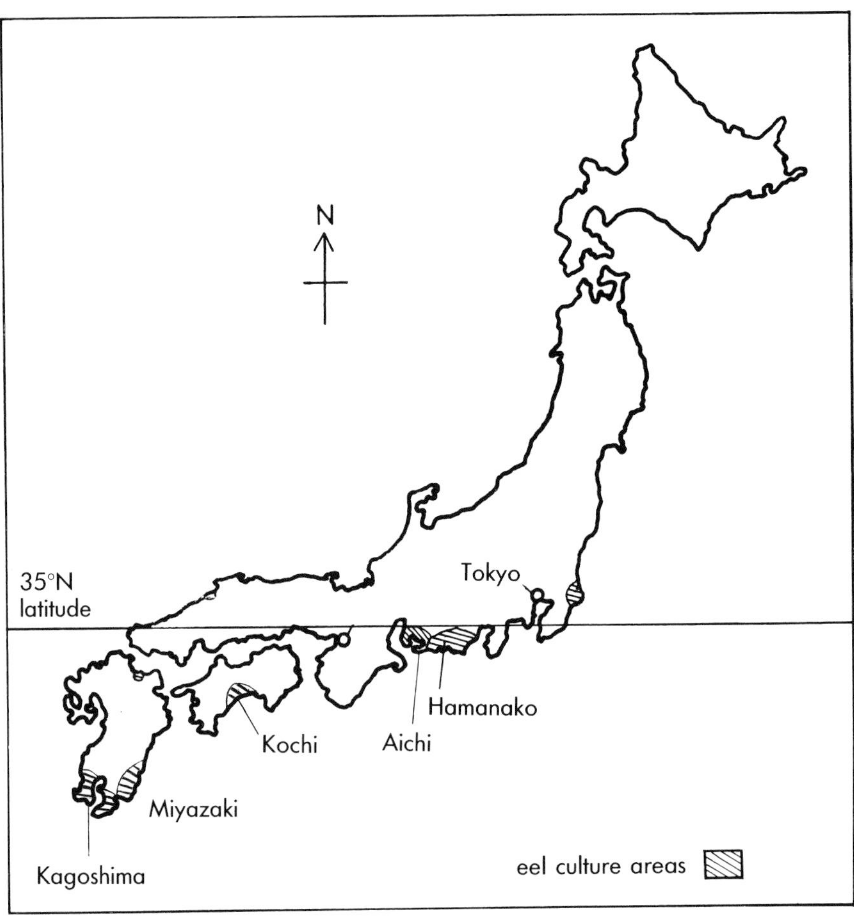

Fig. 6.1 Map of Japan showing the areas where eel farming is carried out.

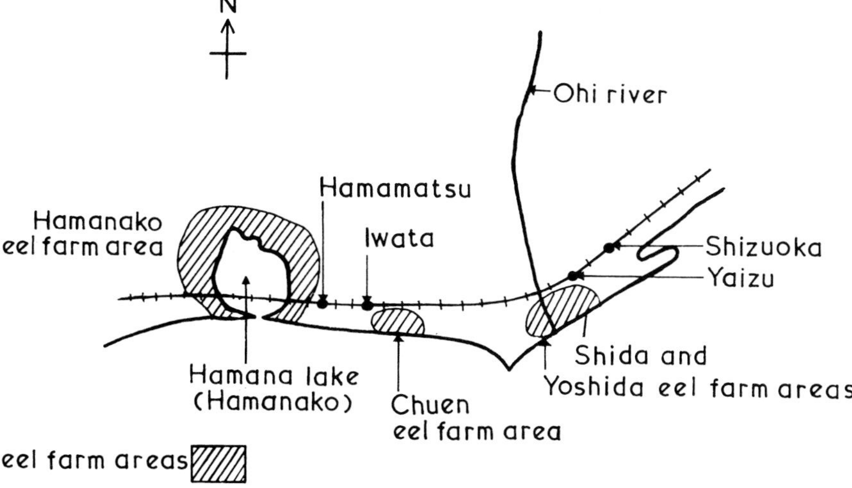

Fig. 6.2 Detailed map of Hamanako lake area, the main centre of eel culture in Japan.

hours in a bath of Oxolinic acid to disinfect them of any adhering bacteria.

The elvers are placed in the elver tank at a density of about 400 g/m^2. Twenty to fifty per cent mortality of elvers may occur during the first month. It is important that weak, stunted or diseased elvers be removed as soon as possible.

After 20–30 days in the first pond, the elvers are caught, sorted into two size groups and transferred to the second elver ponds. All undersized and stunted elvers are separated. In the second pond stocking density is about 100 g/m^2. After a further 20–30 days the elvers, now about 12 cm long, are caught again, sorted by size and transferred to the outdoor fingerling ponds.

In June–July, the fingerlings are caught and put in adult ponds. Many fingerlings are sold to other farmers at this stage. In August–September when 20–30 cm long, the eels are caught and graded by size, and the large and small sized eels are separated into different ponds, in which they over-winter and live until they reach market size at the end of the second summer's growth. The rate of weight increase is shown in Fig. 6.7 and Table 6.3. A typical eel farm is shown in Fig. 6.8.

How long it takes for cultured eels to reach market size is a question often asked. It is difficult to answer because the rate of growth varies between ponds and from farmer to farmer according to the efficiency of his operations. And each year eels grow faster as better techniques are introduced. At present most eels reach 60 g by the end of the first growing season and marketable size of 150–200 g at the end of the second growing season. But even now some farmers in southern

Fig. 6.3 A cluster of greenhouses for eel culture, in Nishimikawa District, Aichi Prefecture, a leading eel-raising district in Japan.

Fig. 6.4 Pumps used for raising water from a borehole.

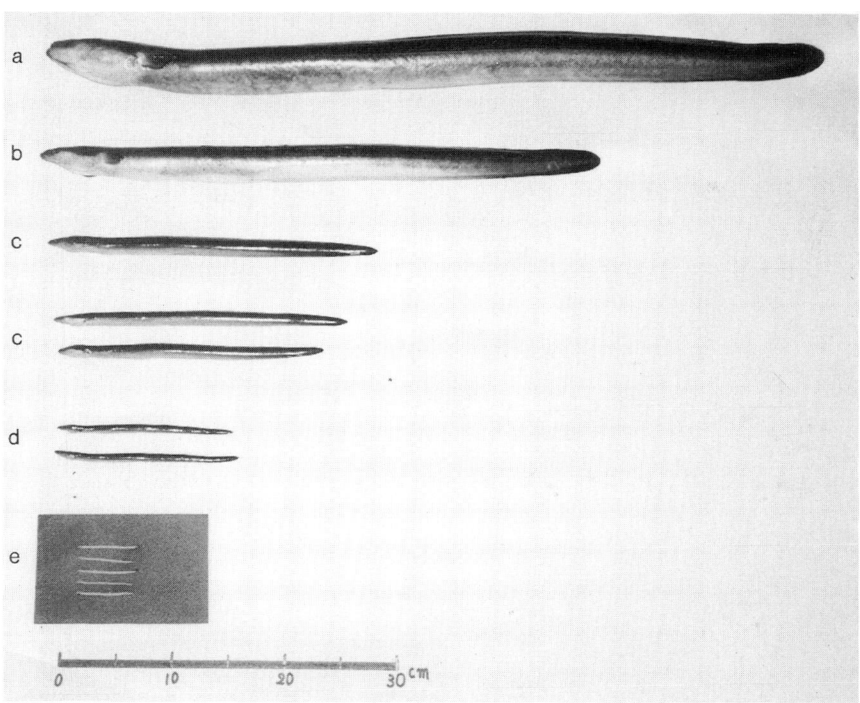

Fig. 6.5 Japanese names and lengths of eels at different sizes and stages of growth.
(a) market size (large) – *Boku*
(b) market size (small or medium) – *Futo*
(c) fingerling – *Genryo*
(d) large elver – *Kuroko*
(e) elver or glass eel – *Shirasu unagi*

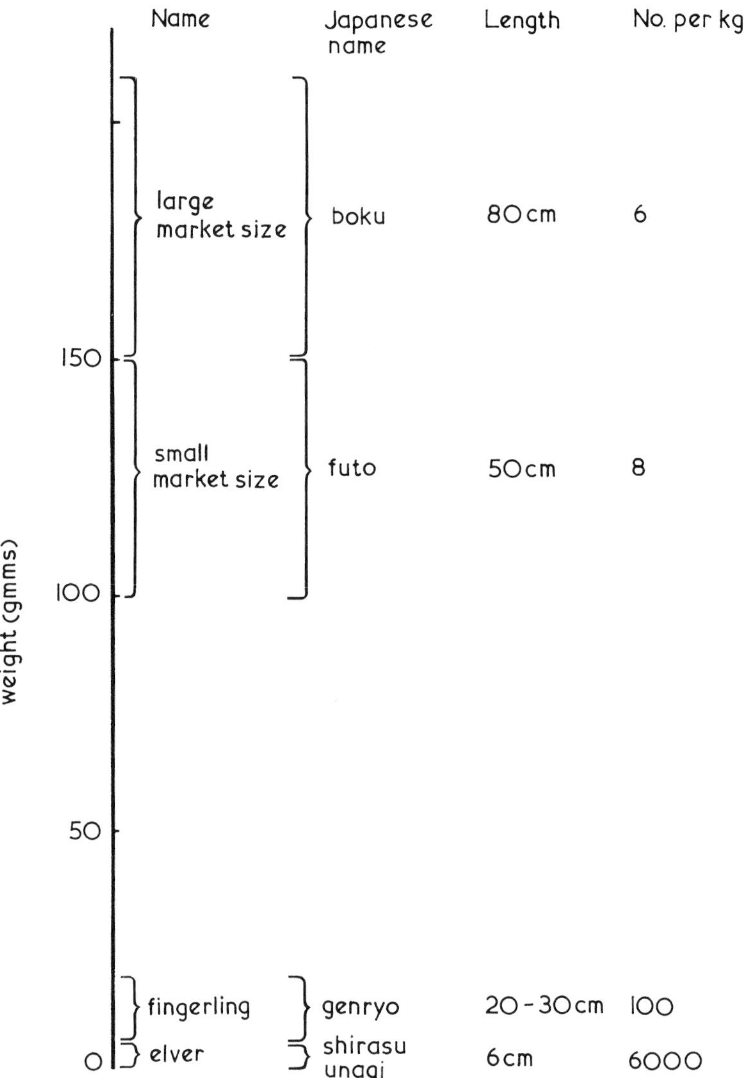

Fig. 6.6 Graph showing names and lengths of eels at different stages of growth. In practice small variations in the figures given may occur.

Japan harvest 120 g eels in November of the year in which the elvers were caught. Many eels are silver stage when they are harvested.

In the second growing season, the young adult eels quit hibernation when the temperature rises above 12°C in April. They are fed once per day in the morning, reach market size in the autumn or early winter and are marketed during the winter months, October to March.

The annual weight budget of eels in a typical 200 m^2 adult eel pond is approximately:

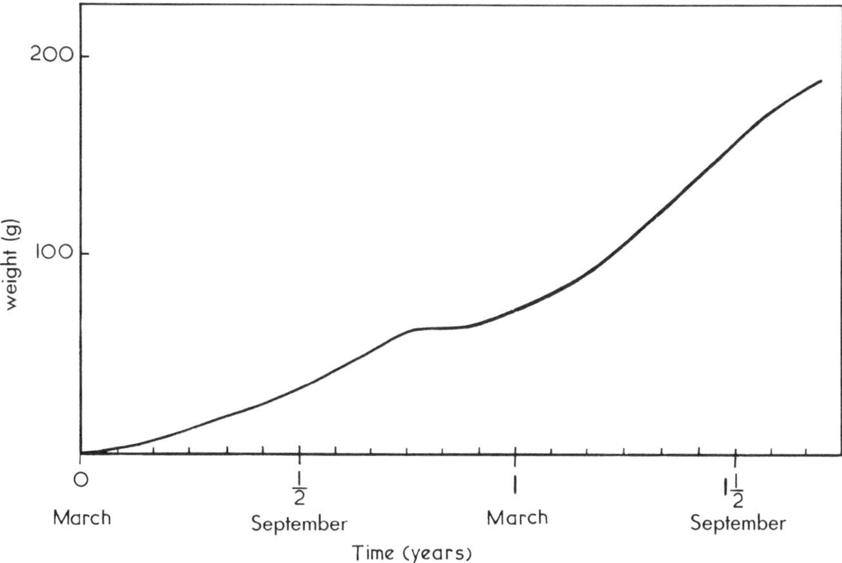

Fig. 6.7 Growth rate of eels in a typical pond. Weight of 60 gm is attained after one growing season, 190 gm in two growing seasons. Both faster and slower rates commonly occur.

Fig. 6.8 View of a typical eel farm in Japan with ponds of about 0.5 ha (a little over 1 acre).

Weight gain:		
Undersized eels carried over from previous year	200 kg	
Fingerling eels stocked in August	100 kg	
Meat weight gain due to feeding	700 kg	
		1000 kg
Weight loss:		
Winter harvest of eels	800 kg	
Undersized eels left to grow	200 kg	
		1000 kg

General rules that will promote good results in eel culture are:

- Keep your eels in the best possible health. Thus careful control of water quality, avoidance of parasites and disease, and avoiding damaging the skin of eels when netting or transferring them will all help.
- Continually sort the eels during their growth, dividing the different sizes so that in any pond the eels are of almost the same size. Separate all undersize or stunted eels. This is one of the main techniques of successful eel-farming.

Given normal mortality, 1 kg of elvers will yield a harvest of 400 kg of market-size eels, i.e. if you desire a harvest of 40 t of eels, you must start off with 100 kg of elvers.

The Japanese have tried out many variations of culture methods, pond designs and feeds during the past 80 years. Rather than describe the successful methods in one chapter, this book now presents the techniques in a step-by-step manner so as to guide persons who live

Table 6.3 Monthly weight increases of cultured eels during their first growing season in Japan

Date	Percentage of year's weight increase	Average temperature of water °C
End of March (new caught elvers)	0	11
End of April	1	16
End of May	3	21
End of June	7	25
End of July	14	28
End of August	24	30
End of September	24	23
End of October	22	19
End of November	4	14
End of December	0	9
TOTAL	100	

in Europe, Australia and New Zealand or any other suitable locality on how to start eel farming.

The success of eel culture in Japan is based on sound business management as well as on good biological techniques. The main divisions of the eel culture industry and their relationships are shown in Fig. 6.9. When eel culture first started, the pioneers caught their own elvers and raised them through to market size. Now the position is much changed.

Elvers are caught at river mouths all around the southern Pacific coasts of Japan but supplies can no longer meet the demand and additional elver supplies are imported by air freight from Europe. Many farmers grow eels only from 20–30 cm fingerling size up to market size. Thus other smaller farms have developed which rear elvers to fingerling size, and then sell their fingerlings to the other farmers.

In each district, eel farmers' co-operatives have long been in existence and all farmers are members. The co-operatives have many useful

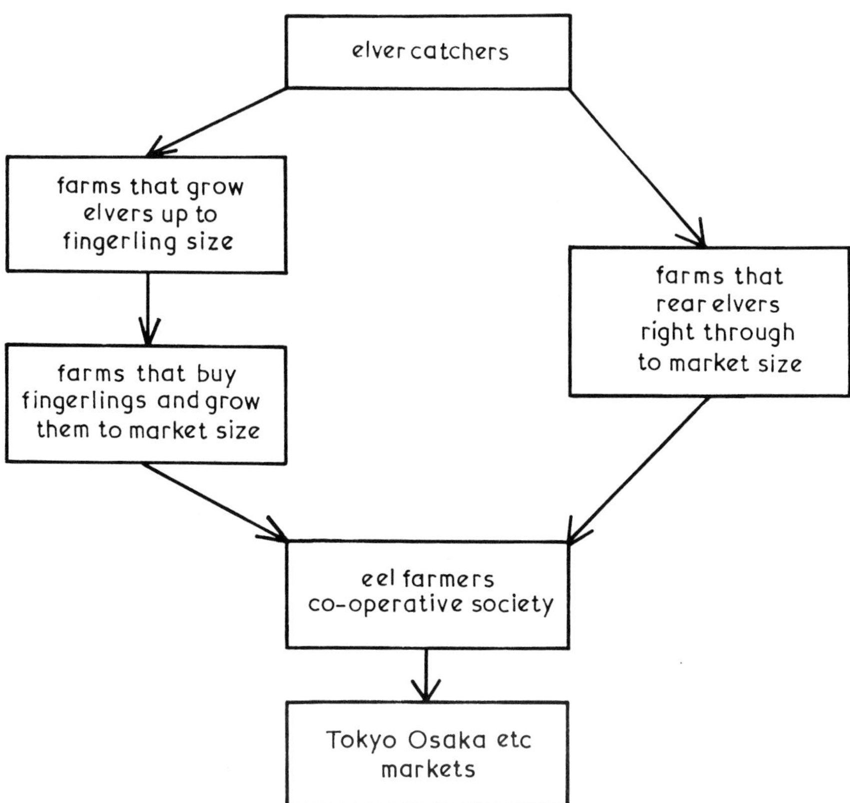

Fig. 6.9 Organization of the eel culture industry in Japan. The establishment of the co-operatives is one of the principal reasons for the success of the industry.

functions, such as owning cold stores, buying artificial feed and fish in bulk at wholesale prices for the farmer members, marketing the harvests of farmers and paying each farmer his share of the total proceeds, negotiating with the Government etc. and assisting members with accounting and legal matters.

Additional to all the private enterprise organizations are well-equipped government research and experimental stations in each area. At many technical colleges and several universities, students can take training courses in aquaculture.

There is one eel farming area about 100 miles north-east of Tokyo, which is within the Pacific Ocean zone. There are no large eel-culture ponds in the estuary of the Tone river. There are very few eel culture ponds in the neighbourhood of Tokyo, but in the areas south of Tokyo there are many more eel culture ponds than in the north.

Table 6.4 Management unit and area in Shizuoka Prefecture in April 1990

District	Management unit number	Greenhouse area (ha)
Shida	54	15
Haibara	201	22
Chuen	18	7
Seien	155	77
Total	428	121

Table 6.5 The tonnage of eels from the areas farmed in the various Japanese prefectures in 1988

Prefecture	Production (t)	Area (1000 m^2)
Aichi	12 064	3 402
Shizuoka	8 654	4 134
Kagoshima	6 367	553
Miyazaki	3 365	592
Kochi	2 945	385
Tokushima	1 506	409
Mie	1 354	390
Oita	735	127
Okinawa	607	67
Kumamoto	273	35
Chiba	197	80
Ehime	150	59
Kanagawa	146	109
Ibaraki	134	21
Gumma	128	4
All Japan	38 625	10 367

The Shizuoka Prefecture and other prefectures are shown in Figs 6.1 and 6.2. Table 6.4 gives the latest details of units and areas in the Shizuoka Prefecture.

The main centres of eel culture in Japan are the Aichi Prefecture and the Shizuoka Prefecture. Table 6.5 gives the tonnage of eels from the areas farmed in the various prefectures during 1987.

7 Choosing a site for an eel farm

A good site for an eel farm must satisfy as many of the following conditions as possible:

- A good supply of water, channelled or pumped from a river or pumped from bore holes. About 450 m^3 per day of water are needed to grow 20 t of eels per year.
- Water may be clear or turbid, but must not be liable to poisonous pollution by insecticides etc. Alkaline or neutral water is best. Acid waters are not suitable for eel culture. Plenty of wild eels living in the water you propose to use is a good sign; it proves its suitable quality.
- Not subject to flooding.
- High enough above adjacent outlet level so that ponds can be emptied by gravity alone.
- Non-porous soil, so that water does not leak out of ponds; sandy clay is best.
- Sunny position to encourage growth of oxygen-producing phytoplankton.
- Open windy position to encourage oxygenation of the surface water by the wind.
- Roads and electricity available.

In Japan, certain flat lands adjacent to rivers and brackish lagoons have been developed into huge eel culture centres. For reasons of both psychology and economy, Japanese eel farmers prefer to locate their farms adjacent to existing farms, and this is certainly beneficial. Since the eel farms are so concentrated, it has been easy for co-operatives to develop. Offering cut price central frozen-feed stores, equipment supply and accounting, subsidiary and marketing services. This concentration also makes it convenient for the government to set up experimental stations in each area and to offer free disease-prevention services etc. to all farmers.

In Europe, flat lands adjacent to big rivers or estuaries would be suitable to develop into eel farms, or, for single farms, sites near disused water mills offer the advantage of a gravity-fed no-cost water supply without pumping costs.

8 Eel farm design and construction

All Japanese ponds have the still water type of irrigation. Figs 8.1 to 8.8 show many details of typical Japanese eel ponds. The trend nowadays is for ponds to become smaller and smaller. At present (1991) cultured eels are accommodated in a sequence of four pond sizes as they grow from elver to market size (Table 8.1).

(1) First elver tank, usually under a greenhouse.
(2) Second elver tank, usually under a greenhouse.
(3) Fingerling pond, greenhouse mainly, outdoors in summer.
(4) Adult eel pond, greenhouse mainly, outdoors in summer.

ELVER TANKS

Elver tanks are usually built under greenhouses and are heated to above 25°C by thermostatically-controlled electric immersion heaters. This heating gives the elvers a fast start in the spring, a technique also used in trout and oyster farming. The first elver tank into which the wild-caught elvers are placed and reared for a month is usually

Fig. 8.1 Bulldozing flat land to make a pond.

Fig. 8.2 A newly dug pond before the banks are lined.

Fig. 8.3 Lining pond banks with grooved concrete slats and posts.

circular, 5 m diameter by 60 cm deep and made of concrete. Water is sprayed onto the surface from angled jets, creating a circular whirlpool current. The water drains away through a central pipe. The second elver tanks raise the elvers from 8 cm up to 12 cm size and are about 30–100 m^2 area and 1 m deep. Both types of elver tanks are aerated

Fig. 8.4 A completed bank of grooved concrete slats and posts.

Fig. 8.5 Preparing to build boulder and concrete pond banks. Note the drainage ditch outside the bank to the right.

by bubblers as well as by the surface jets of incoming water.

It is a good idea to obtain the water supply for elver tanks from a bore hole or well. This guarantees a pure supply for the delicate elvers. Larger sizes of eels are more robust and can tolerate the variations of water quality that may occur sometimes in river water.

Fig. 8.6 Building a pond bank with boulders.

Fig. 8.7 Empty pond showing mud bottom and solid concrete banks.

FINGERLING PONDS

Fingerling ponds are usually rectangular, of about 200–300 m^2 in area, and 1 m deep with a mud bottom. To prevent the escape of

Fig. 8.8 A solid concrete bank.

small eels on rainy nights when rain trickles into the outdoor pond, it is desirable for the concrete pond edge to have an overhanging lip for elvers and fingerling eels. Above the length of 20 cm eels do not try to leave the pond, and so in adult eel ponds no overhanging lip is necessary.

Adult eel ponds are rectangular or square and were formerly made between 5000 and 20 000 m^2 (0.5 and 2.0 ha) in area, as shown in Figs 8.1 to 8.8. Recently the size has become greatly reduced to only about 500–1000 m^2 in area and may be further reduced in future. Increased sophistication of management is needed in the care of small ponds.

By the running water culture method and the use of efficient drainage with the water at temperatures of about 28°C, marketable-sized eels to a total weight of 1.5–2.5 t (10 kg/m^2) can be produced in a concrete pond, 1 m deep and 15–18 m in diameter with a concrete bottom.

In Japan and Taiwan at the present time, adult eels are raised in several types of ponds dating from different periods of construction. Large ponds, in some cases completely stagnant, are used in some areas where the water supply is small but where flat land is plentiful and cheap.

A drainage ditch about 1 m wide and with its bottom at a lower level than the pond bottom, is excavated adjacent to the exit sluice of each pond to carry away discharged water.

Table 8.1 Pond space required to yield 20 t of market size eels per year. 50 kg of elvers are needed to produce 20 t of market size eels.

A water supply of about 450 m^3 (100 000 gallons) per day is needed.

Size of eel stocked	Weight stocked start finish	Weight & length of average eels start finish	Stocking density start finish	Pond area needed m^2	Number and size of ponds
Elvers	50 kg	0.16 gm 6 cm	0.4 kg/m^2	125 m^2	6 ponds 20 m^2 (5 m diameter) size,
	150 kg	0.5 gm 8 cm	1.2 kg/m^2		running water
Older elvers	150 kg	0.5 gm 8 cm	0.5 kg/m^2	300 m^2	8 ponds 40 m^2 (8 × 5 m) size
	400 kg	1.3 gm 12 cm	1.6 kg/m^2		running water
Fingerlings	0.4 tons	1.3 gm 12 cm	0.4 kg/m^2	800 m^2	10 ponds 80 m^2 (10 × 8 m) size
	2 tons	6.5 gm 20 cm	2.0 kg/m^2		still water
Adult eels	2 tons	6.5 gm 20 cm	0.4 kg/m^2	5 000 m^2	25 ponds 200 m^2 (14 × 14 m) size
	20 tons	190 gm 70 cm	4.0 kg/m^2		still water

CONSTRUCTION

Ponds are usually constructed by excavating the bottom of the pond using a bulldozer, piling the excavated earth around the pond to form banks, and then facing the inward side of the banks with concrete slats or cemented stones to prevent them from being eroded by waves. Banks between ponds are usually about 2 m wide at pond bottom level and 1.4 m wide on top. Table 8.2 shows the costs of constructing two large ponds in Japan in 1987.

Each pond has one main water inlet and one main water outlet. The inlet is either a pipe bringing in pumped water or a sluice leading water in from a river. If the incoming water can be poured into the pond from a height of 60–100 cm extra oxygenation will occur. Such pouring from a height can easily be arranged when piped water is used.

The bottom of the pond is built to slope slightly downwards toward the outlet sluice and is shallowest, about 80 cm deep, near the water inlet and deepest, about 120 cm, near the water outlet. This allows the pond to be completely drained by simply opening the outlet sluice.

Figs 8.9 to 8.14 show ponds during and after construction. Sluice gates at the outlet of each pond consist of a concrete lined gap in the pond wall bearing slots for three different kinds of doors (Fig. 8.12). To cater for the various operations of pond management the three types of doors are vital and they fulfil the following functions:

- 1st slot: board door, permits the pond operator to choose whether surface or bottom water layers of the pond are discharged.

Table 8.2 Cost of constructing eel ponds in Japan 1987
Prices in Yen, 242 Yen = £1 (present 1990 rate: ¥ 252 = £1)

	5 000 m² pond lined with stones and cement	12 000 m² pond lined with concrete slats
Work of bulldozer to excavate pond	1 710 000	2 400 000
Stones and cement	600 000	
Concrete slats and posts		900 000
Electrical wiring	375 000	450 000
Digging well and laying water distribution pipes	750 000	675 000
Vertical pumps	90 000 (one pump)	270 000 (three pumps)
Paddle splasher machines	90 000 (one splasher)	360 000 (four splashers)
Feed store	600 000	675 000
	¥4 215 000	¥5 730 000
	£ 17 400	£ 24 000

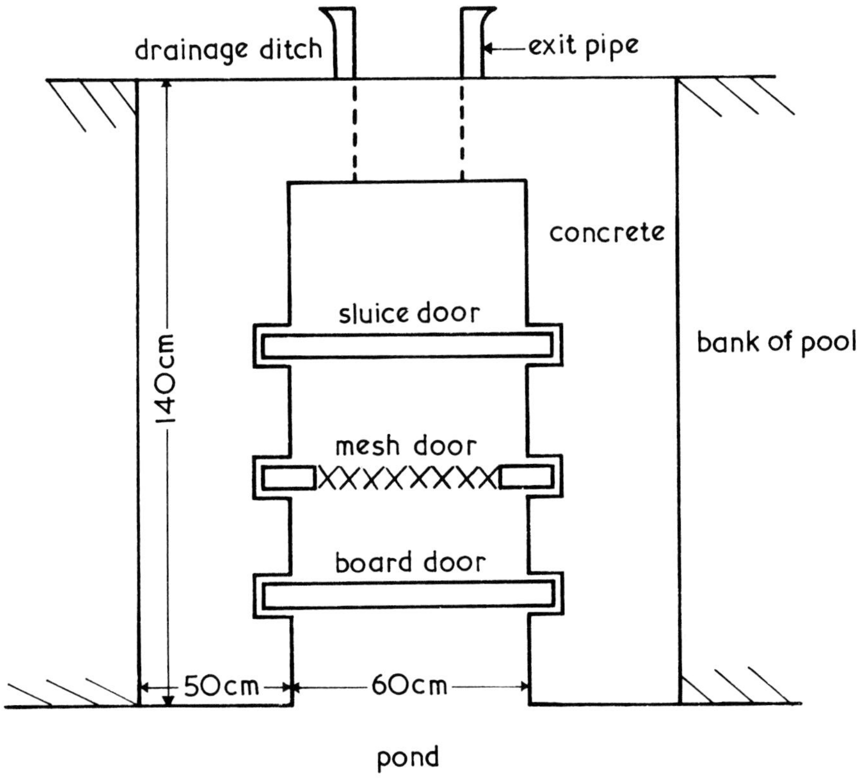

Fig. 8.9 Top view of a sluice gate showing arrangement of board, mesh and sluice doors in their slots.

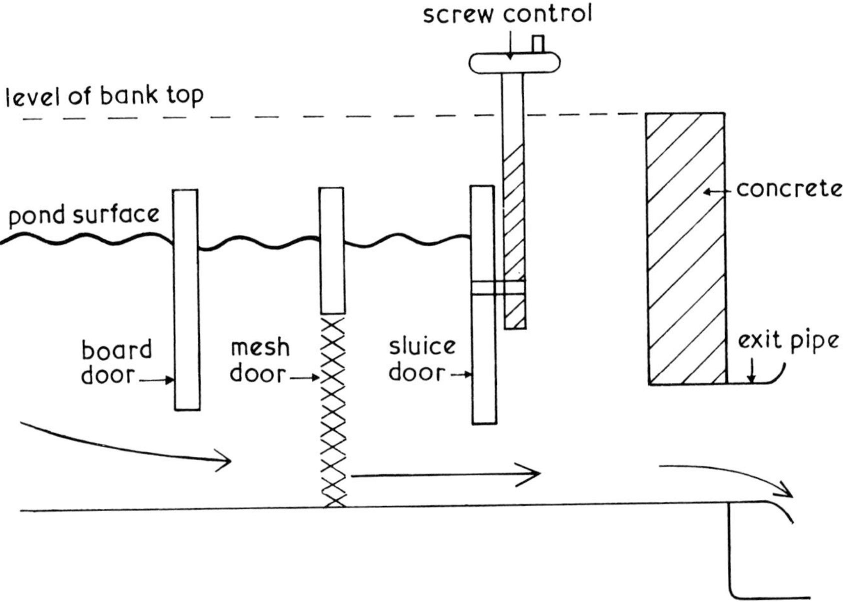

Fig. 8.10 Vertical section of the sluice gate showing the three doors. The board door is set so that the water from the lower layers of the pond can be drained.

Fig. 8.11 Building the exit sluice of a pond.

Fig. 8.12 The completed exit sluice showing the three doors.

Fig. 8.13 Drainage ditch lined with concreted boulders.

Fig. 8.14 The exit sluice discharges into a drainage ditch by way of a large diameter concrete pipe. By tying a net bag over the lip of the pipe end, eels can be harvested by draining the pond.

- 2nd slot: mesh door, prevents the escape of eels.
- 3rd slot: sluice door, controls the amount of water leaving the pond.

The size, strength etc. of a given inlet or outlet sluice gate will vary according to the volume of water flow with which it has to cope.

On the outside of pond outlet sluices the water is discharged through a big pipe into the drainage ditch. By tying a bag net onto the end of this pipe eels can be harvested when the pond is being drained (Fig. 8.14).

An alternative method of permitting water to escape from ponds is by the use of 20 cm diameter plastic tube siphons set across the pond bank (Figs 8.15 to 8.17). A net cylinder around the inlet end prevents fish escaping; a hand pump applied to an air valve at the top is used to exhaust air from the pipe and set the water siphoning over. The outlet end of the pipe discharges into the drainage ditch.

Electric wires supplying the pumps etc. should be buried in the pond banks if possible, or hung from poles. It is desirable that electric switches controlling pumps and aerators etc. should be installed all together at the main watchman's hut.

To prevent the banks from collapsing, they must be faced in some durable way. The most common method is to use slats and grooved posts of concrete (Fig. 8.4). The grooved posts are set at 1.3 m intervals and the slats, which are exactly 1.3 m long, 3 cm thick and 35 cm wide, are slotted into the grooves to make a solid wall. Boulders

Fig. 8.15 Siphon draining water from the pond.

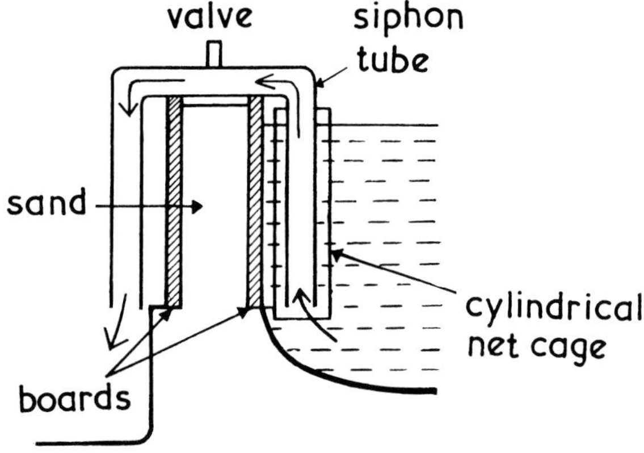

Fig. 8.16 Siphons used as a water outlet.

Fig. 8.17 A siphon being used to drain one pond into an adjacent empty pond. Pumps are for raising water up from a borehole.

cemented to form a wall are also used but this requires more labour and expense. The facing of the banks should extend from 30 cm below the level of the pond bottom to 50 cm above the pond surface.

Ponds for adults and fingerlings have a special feeding place incorporated in one bank of the pond (Figs 8.18 to 8.21). The feeding point should be sited on the side of the pond opposite to the source of the

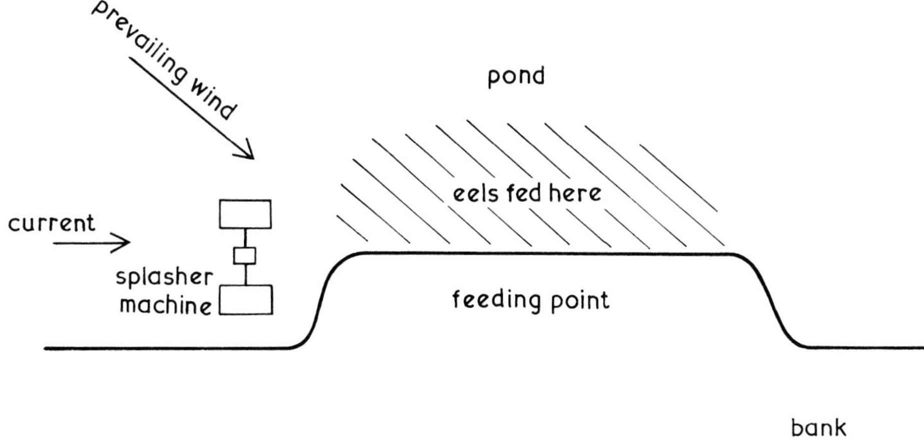

Fig. 8.18 It is wise to establish a customary feeding point in the pond. This should be sited at a position where the oxygen level is high as this encourages the eels to eat well.

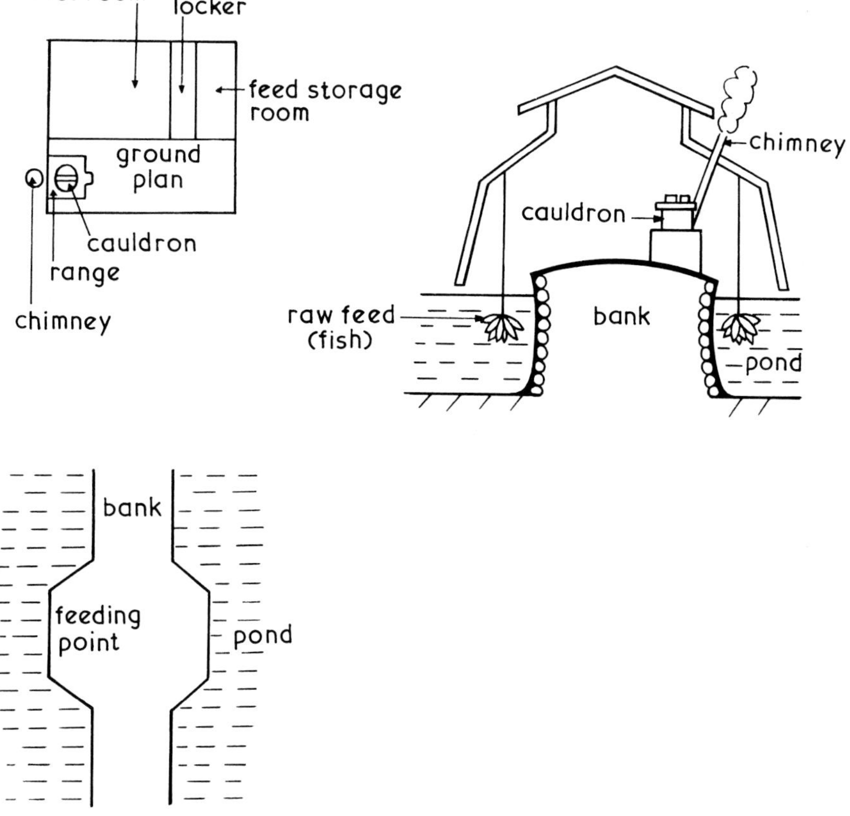

Fig. 8.19 A hut and cooking equipment are sometimes provided at the feeding point.

Fig. 8.20 A shaded feeding place.

Fig. 8.21 Another type of a shaded feeding point.

prevailing wind. Such waves increase oxygenation of the water, which in turn increases the eels' appetites. The feeding point consists of a built-out bulge from the main bank, which presumably deflects the water flow of the pond and thus makes it easy for eels to locate it. In some cases a shed or platform of some sort is erected so as to shade the feeding point because eels like darkness.

Many farmers construct a 'resting corner' in the corner of adult eel ponds that are larger than 600 m^2 (Figs 8.22 to 8.24). The resting corner is really a running water pond linked to the main still water pond. The resting corner has a water inlet and outlet, a splasher-aerator machine and a vertical pump, and has a water circulation that

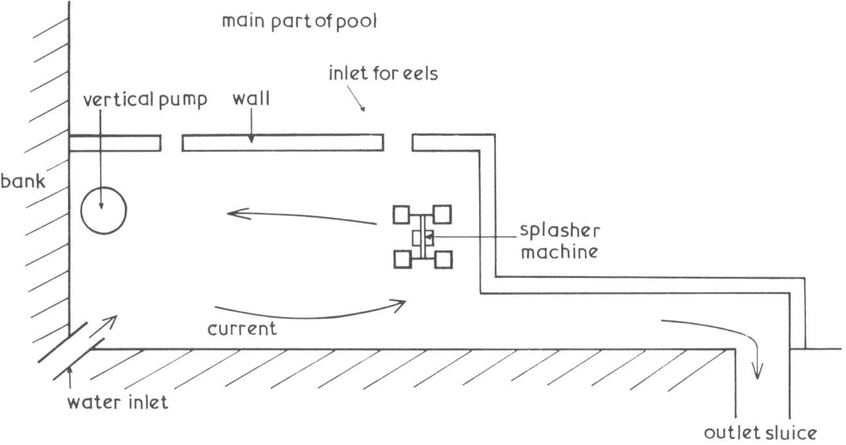

Fig. 8.22 A typical example of 'resting corner' in an eel pond.

Fig. 8.23 The resting corner in a pond.

Fig. 8.24 Another design of resting corner.

is almost separate from the water in the main part of the pond and is of particularly high oxygen level and purity. It connects with the main body of the pond by a small opening only. Thus, if water quality is bad in the main body of the pond, such as at night when the oxygen level falls, the eels enter the resting corner and find better conditions. Likewise, when eels are netted during harvesting, the dense bag of eels is dragged to the resting corner for sorting etc., and the higher oxygen supply in the corner saves them from dying of lack of oxygen. A small opening in the wall dividing the resting corner from the main part of the pond allows eels to enter or leave as they wish.

A typical resting corner occupies two per cent of the area of these large ponds. At night, extra water can be pumped into the corner, the aerators switched on and oxygen thus kept high until morning.

9 Water quality and its control

As air is to humans, so is pond water to the eel. If the quality of the pond water is good, eels will be healthy and grow fast. If it is poor, they will be unhealthy and grow poorly.

The key to good water quality is good circulation, so the shape of ponds and the positioning of inlets, outlet sluices and paddle machines must ensure that all the water of the pond circulates freely. If any stagnant water exists it will be sure to cause trouble.

The characteristics of the water that are most important are the oxygen content, pH, nitrate, temperature, and the amount of phytoplankton. (For details of fast-flow types of ponds which have no phytoplankton see Chapter 5.) The water in a pond is subject to many influences, which vary during each cycle of 24 hours, from week to week and from season to season. The feeding and breathing of the eels, the growth of plankton, the running in of new water and running out of old water, all influence the water quality. But most of all it is the growth of the phytoplankton in relation to the rising and setting of the sun each day which affects water quality.

The normal daily variation of oxygen, pH and temperature in a pool is shown in Fig. 9.1. Oxygen ranges between 1 and 10 ml/litre and pH between 7 and 9. Nitrate in healthy ponds varies from 0 to 100 mg/litre during the course of each phytoplankton bloom and decline.

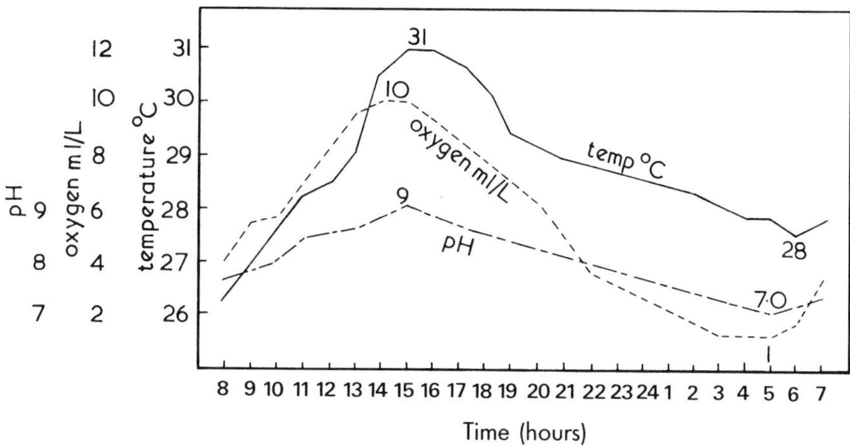

Fig. 9.1 The normal daily cycle of water quality which occurs in a typical eel pond in summer.

Typical eel ponds are a thick soupy green colour in summer, and this growth of phytoplankton is encouraged because by day the phytoplankton produces much oxygen which augments the natural supply in the water and allows large weights of eels to live in the pond. At night, however, the phytoplankton uses up oxygen and on hot summer nights the oxygen levels of ponds frequently drop below the level at which eels can breathe. Certain kinds of phytoplankton are dangerous; e.g. *Anabaena* blooms reduce the appetites of eels greatly. Zooplankton is not desirable in eel ponds, because blooms can cause severe oxygen depletion. A healthy pond contains phytoplankton and zooplankton in proportions about 98:2 by weight.

Phytoplankton does not exist in a steady stage of abundance but goes through cycles of growth, reproduction, spore production and death. These occur about every 40 days in summer and the colour of the pond and its water quality change according to the stage of the cycle (Fig. 9.2). A bloom is the name given to the sudden appearance in a pond of dense masses of plankton. When a bloom or a sudden death of plankton occurs, the chemical balance of the pond is temporarily upset and there is danger of oxygen depletion. The transparency of a pond is a measure of the density of phytoplankton in the water and should be maintained in such a state that a white disc lowered into the pond is still visible when 10–40 cm below the surface.

Oxygen is the most critical water characteristic. At oxygen levels below 1 ml/litre eels cannot breathe and rise gasping to the surface. A good eel farmer frequently checks his ponds, especially at night, to see whether any fish are swimming gasping near the surface. If he sees any, it is a sure sign that urgent action is needed to increase the oxygen supply. It is a good idea to test the oxygen content of the

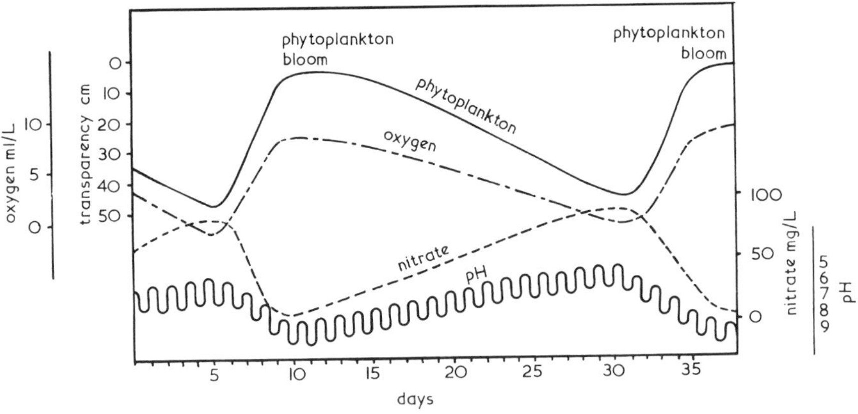

Fig. 9.2 The phytoplanktonic growth cycle and its effect on water quality. During the bloom the pond becomes a thick soupy green, nitrate is at a minimum and the pH is alkaline. According to the species of phytoplankton which grow in the pond and the weather, blooms occur approximately every 40 days during the summer.

bottom water of the pond also and use vertical pumps to raise it if necessary (Fig. 9.3).

Oxygen deficiency regularly occurs under the following conditions:

- on summer nights between 8 pm to 6 am, during which time the phytoplankton use up much oxygen;
- when many eels are concentrated together, such as when harvested in a net;
- when deaths of phytoplankton occur, or when harmful species of phytoplankton multiply in the pond.

There are several methods to improve the water quality in eel ponds:

- Let more water than usual flow into the pond, thus increasing the rate of water exchange and letting foul water pass out of the pond.
- Switch on paddle splasher machines. These splash water into the air, thus oxygenating it and improving horizontal circulation in the pond (Figs 9.4, 9.5, 9.6).
- Switch on vertical pumps. These draw water from the lower levels of the pond and shoot it out over the surface, thus oxygenating the water, improving vertical circulation in the pond and preventing accumulation of foul water on the pond bottom (Figs 9.7, 9.8).
- Add various chemicals to the water such as CaO to reduce acidity, iron oxide to remove hydrogen sulphide.
- In extreme cases a pond must be emptied, dried out, disinfected and refilled with new water.

A great range of models of paddle splasher machines and vertical pumps exist in Japan and a selection of modern ones is shown in Figs 9.4 to 9.9.

The modern concept of eel culture is to grow as many eels in as small a volume of water as possible. This causes the chemical balance of

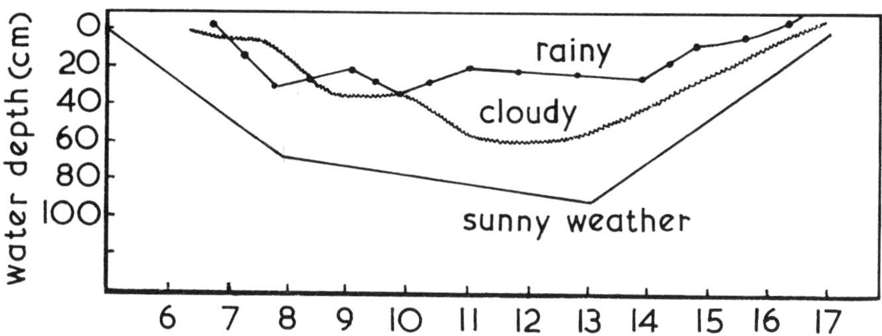

Fig. 9.3 Depth at which depletion of oxygen must be compensated for by bringing in fresh water or by agitation according to weather (see gauge). X-axis shows time of day (24-hour clock).

Fig. 9.4 A typical paddle splasher machine for adding oxygen to the water.

Fig. 9.5 A special paddle splasher having two pairs of paddles, one pair deep and the other shallow.

Fig. 9.6 A simple paddle splasher driven from the bank.

Fig. 9.7 A vertical pump discharging onto a splash board to give added oxygenation.

Fig. 9.8 A vertical pump in the resting corner of a pond.

Fig. 9.9 A fountain aerator which is sometimes used is spectacular but not economical.

the pond to be very precarious and demands sophisticated quality control. Oxygen deficiency arises regularly during summer and on many farms the paddle splasher machines and vertical pumps are run continuously during warm weather.

Each morning and evening farmers test the oxygen, pH, nitrate, plankton etc. levels in each pond and switch on the aerator machines or adjust the rate of water inflow to bring about optimum water quality. Some simple equipment and a microscope are needed for doing the tests.

Most Japanese eel farmers sprinkle powdered lime and ferric oxide (Fe_2O_3, trade name Manken) on the surface of ponds every 1–2 weeks. The former aids decomposition of organic waste, excretions etc. in the pond and the latter aids phytoplankton growth and destroys any harmful sulphides in the water by combining with them to form an insoluble precipitate of ferrous sulphide.

Rotifers of the species *Brachionus plicatilis* are among the most dangerous zooplanktonic animals that occur in eel ponds. These minute swimming animals feed on phytoplankton and use up much oxygen, causing wholesale asphyxiation of eels. When microscopic study of a water sample reveals more than three rotifers in the field of view the only course of action is to completely change the water in the pond as fast as possible.

Plagues of water fleas, *Cladocera*, sometimes occur in ponds, but can easily be killed by sprinkling Masoten on the water.

10 The feeding of eels

In Japanese farms artificial feed and raw fish are used to feed eels and the use of artificial feed is rapidly increasing. At present about 70 per cent of farmers use artificial feed and 30 per cent use raw fish. In Japan feed amounts to about 30 per cent of the production cost of cultured eels.

The conversion ratio of raw fish and artificial feed is shown in Table 10.1. Much less artificial feed than raw fish is needed to bring about the same increase in eel weight. The apparently great difference is due to the fact that artificial feed is dry. The conversion ratio of both types of feed is higher in warm temperatures than in cold, and among young than among old eels.

Artificial feed is convenient to use because it does not require cold storage and needs little space. It is produced in powder form and is mixed with water and 5–10 per cent of vitamin oil to make a paste (Figs 10.1, 10.2, 10.3). This paste is placed on a mesh tray in the eel pond and the eels go through the mesh or climb on to the tray to get the food (Fig. 10.4). Uneaten paste remains on the tray and is lifted out of the pond, thus does not foul the pond. Artificial feed is made chiefly of fish meal with added carbohydrate and consists of about 52 per cent protein, 24 per cent carbohydrate, 10 per cent water, 4 per cent fat and 10 per cent ash. Artificial feed caused enlargement of eel livers when first introduced but its composition is now amended to avoid this. The percentage of vitamin oil added to the feed is varied according to the water temperature; below 18°C about 5 per cent by weight is added, above 18°C about 10 per cent is added.

Raw fish used are mackerel (*Scomber*), launce (*Ammodytes*), Pacific saury (*Cololabis*), atka mackerel (*Pleurogrammus*), horse mackerel (*Trachurus*), sardine (*Sardinops*) and anchovy (*Engraulis*). These fish are usually bought in frozen blocks, then thawed overnight and

Table 10.1 Conversion ratio of artificial feed and raw fish

Type of Feed	Conversion ratio	Meaning
Artificial feed	1.4:1	1.4 kg of feed (dry weight) must be eaten by the eels to bring about a weight increase of 1 kg
Raw fish	7:1	7 kg of feed must be eaten by the eels to bring about a weight increase of 1 kg

Fig. 10.1 A mincing machine for preparing the food.

Fig. 10.2 A machine for mixing artificial feed. The ingredients may vary slightly from time to time.

Fig. 10.3 Artificial feed for eels being mixed with water and vitamin oil. Photo G. Williamson.

Fig. 10.4 Eels eating artificial feed from a tray lowered into the pond. Once they have ceased feeding it is lifted out to avoid scraps fouling the water.

fed whole to the eels. Offal is not used because it has poor nutritive content and tends to cause outbreaks of disease. The chemical analysis of various fish is given in Table 10.2.

It is vital that no unnecessary organic debris be allowed to fall into ponds. Thus, when feeding eels, the food is not just thrown into the pond. Instead, a special feeding point is established at one side of each pond and the eels learn to come there each feeding time. Artificial feed is lowered on a perforated tray (Fig. 10.4). Raw fish are threaded through their eyes on a piece of wire (Fig. 10.5), dipped in a tureen of boiling water for a minute to soften the skin and then lowered into the pond (Fig. 10.6). After the eels have eaten off the flesh (an orgy lasting less than five minutes!) the wire is raised, bringing out the head and backbone of each fish, and these are thrown away (Fig. 10.5). The only organic matter that enters the pond is what the eels actually eat.

Eels are fed once only per day, about 08.00 to 10.00, and in the warmest part of the summer the quantity of feed given should be about ten per cent of the body weight of all the eels in the pond. If more than this amount of food is given conversion efficiency decreases and money is wasted. In cool months feed is reduced and during winter no food at all is given.

The quantity of food consumed on a typical eel farm each month is shown in Table 10.3. In spring little food is used and then during the summer consumption rapidly increases for two separate reasons. Firstly, the warmer temperatures encourage eels to eat much, and secondly, the total weight of eels in the ponds steadily increases as they grow.

Naturally the appetite of eels in the ponds also steadily increases as they grow. They eat most on clear, windy, dry days and eat least on rainy, cloudy, calm days.

A few years ago, harvests were sometimes found to contain a preponderance of small eels and on examination these were found to consist mainly of males, which grow to much smaller sizes than females. Eels are unusual in that their sex is not finally fixed until

Table 10.2 Composition of raw fish

Species	Composition (per cent)			
	Water	Protein	Fat	Carbohydrate
Mackerel (*Scomber*)	76	18	4	0.7
Atka mackerel (*Pleurogrammus*)	77	17	5	0.2
Pacific saury (*Cololabis*)	70	20	8.4	0.2
Sardine (*Sardinops*)	75	17	6	0.8
Horse mackerel (*Trachurus*)	75	20	3	0.7
Eel (*Anguilla*)	61	20	18	0.3

Fig. 10.5 The heads and skeletons of mackerel which alone remain after the eels have fed. Threaded through the eyes, these too are removed to avoid fouling the water.

Fig. 10.6 Eels ravenously eating mackerel, which is first dipped for a few minutes in boiling water to soften the skin.

Table 10.3 Monthly consumption of food on a typical eel farm

	Percentage of weight of food eaten annually	Average temperature of water °C
January	0	7
February	0	5
March	1	11
April	2	16
May	3	21
June	7	25
July	15	28
August	24	30
September	28	23
October	18	19
November	2	14
December	0	9
TOTAL	100	

they are over 30 cm long and the above phenomenon is thought to be caused by the crowded conditions of culture somehow influencing more than the normal percentage of eels to become male. The remedy is to mix some hormone with the food of eels, which tends to make them all grow to be females.

Elvers do not take food for the first few days after capture and will only gradually develop an interest in feeding. First feeding of newly caught elvers is done just after dark, using a bright electric bulb suspended over the water to attract the elvers to the feeding point. Care must be taken not to put in the pond more food than the elvers will eat or it will decay. The usual method is to put the feed paste on a gauze tray and lower the tray just beneath the surface at the edge of the elver tank. Let the elvers eat their fill, then remove the tray and any uneaten feed still in it. Below 13°C, elvers do not feed, thus elver ponds must, artificially or naturally, have temperatures above this value.

Elvers are fed a sequence of diets as below:

- Weeks 1 and 2: minced bivalve flesh or minced earthworm flesh.
- Weeks 3 and 4: minced fish flesh/*Tubifex* or similar worms.
- Weeks 5 to 10: minced fish flesh or artificial feed.

New artificial feeds are being developed acceptable to elvers.

The elvers are fed all day, from 08.00 to 15.00 and eat about 25 per cent of their body weight of food per day.

It is not possible to give elvers or large eels one type of food one day and another type the next day; all changes of diet must be introduced gradually by mixing in increasing proportions of the new type of food over a period of about a week.

THE BEHAVIOUR OF ELVERS

Keep some elvers in an aquarium. This is good advice to all eel farmers, because in an aquarium you can actually see what the elvers do, what they like and dislike and how they behave. After raising elvers in an aquarium you have much useful knowledge to guide how you treat the larger numbers of elvers you wish to culture in tanks, where you cannot observe them closely. The phases of behaviour of elvers are as follows.

Newly caught up to the 4th day

The elvers hide in crevices, under stones etc., emerge only at night (when they are attracted by a light) and show no interest in food.

4th–10th day

The elvers hide in crevices, become grey in colour and begin to feed. The best food is mashed earthworms. Dig fresh garden worms, on a wooden block cut them fine with a razor blade, then grind into a paste using the back of a spoon. At first when mashed worm meat is dropped in the aquarium, no elvers take any notice of it unless they actually bump into it by accident. When an elver does this it starts to make feeble bites at the bottom, biting surrounding stones and finally by accident bites the meat fragment. It takes the elvers an hour or more to eat their feed, given once daily.

10th–20th day

The elvers swim actively for about half their time, hide in crevices, among weeds or bury themselves in the gravel with just their heads sticking out the other half of the time. They are darker in colour but their bellies are more or less transparent. After feeding, it is easy to see the food in their stomachs. They feed twice daily at 09.00 and 17.00 hours and the food is always placed in the same corner of the aquarium. Elvers now quickly detect the presence of food and are excited by the smell, but are slow to locate exactly where the food is. As the current diffuses the scent of the food around the aquarium, the elvers suddenly start to swim vigorously or prowl searching along the bottom. They locate the food entirely by smell (blind eels kept in aquariums have lived many years) and grab it greedily when they find it.

21st–30th day

The elvers are now noticeably growing and it can be seen that some grow faster than others. They are now used to their feeding routine, and as soon as the food is put in the aquarium, they quickly smell it and come to the place and greedily grab the food. Each feeding session is over in ten minutes.

After the 30th day

When your elvers get this far, you are well on the way to success. They become ever greedier and there are four vital aims which one must have.

(1) Keep the water crystal clear.
(2) Give the elvers as much food as they can take; they therefore grow quickly.
(3) Never let surplus uneaten food get into the tank or it will poison the water and cause disease. Thus food is not thrown into the pool but is offered on a tray suspended just under the surface. When elvers have had enough, any surplus is lifted out of the tank.
(4) Throw out stunted or undersize elvers whenever they are detected and periodically separate the remainder into fast and slow growers and raise these two groups separately.

Good feeding practice in the early feeding period up to six weeks of age is of great importance as upon the progress in this period depends the final success of the operation. This is in line with the recognised feeding pattern of young life – progress in the early stages lays solid foundation for steady progress later. In one experiment the writer (Mr Usui) achieved in four months a weight of 10 000 grammes for one batch of elvers compared with five months taken for a comparable batch to achieve a somewhat similar weight by a different feeding programme.

11 Diseases and parasites

Eels are attacked by a considerable number of diseases and parasites, the main ones of which are listed in Table 11.1. Many of the diseases are temperature-dependent, i.e. they flourish only within a certain range of water temperatures. Apart from the listed attackers, eels can suffer from malnutrition due to an unbalanced diet.

Prevention is better than cure, so the best advice to eel farmers is to ensure that diseases never get a chance to strike in the first place.

The main methods of preventing diseases and parasites of eels are:

- Ensure that ponds always have a pure and plentiful water supply.
- Dry and sterilize ponds after each harvesting. On dried out and on full ponds scatter lime and ferric oxide to decompose organic wastes on the bottom.
- Give good, balanced diet, sometimes with added anti-bacterial drug etc.
- Avoid damaging the skins of eels when they are handled.
- During the last week of the autumn in which feed is given and during the first week of feeding in spring, add anti-bacterial drugs and vitamin E to the feed.
- Kill intermediate hosts of parasites by adding suitable poison to the pond water.
- Remove dead eels quickly from the pond and burn the bodies.

To cure diseased eels is difficult. It is not practicable to catch eels and transfer them into a medicinal dip. The best treatment when eels are found to be diseased or parasitized is as follows:

- Add medicine to the food of the eels.
- Add medicine to the pond water.
- Pump salt water into the pond (in certain cases).

Knowledge of most diseases is rudimentary at present, and in some cases all a farmer can do is hope for luck, isolate any ponds in which disease occurs, remove eels that die and hope that some survive. In serious epidemics all the eels in a pond may die or may have to be killed, the pond dried and sterilized and a fresh start made with new fingerlings.

FUNGUS DISEASE (SAPROLEGNIASIS, *WATAMAKURI-BYO*)

This is the most serious disease because no effective cure has yet been found (Figs 11.1, 11.2, 11.3). In 1966 and 1968 this disease

Table 11.1 Diseases of Eels

Name	Japanese name	Illustration	Symptoms	Cause	Cure
Fungus disease (Saprolegniasis)	*Watakamuri-byo*	Figs 11.1, 11.2, 11.3	White fungal growths on outside of body	Fungus (*Saprolegnia parasitica*)	Add anti-bacterial drugs to feed; put malachite green or methylene blue in pond water
Red disease (Aeromoniasis)	*Hireaka-byo*	Fig. 11.4	Red colour on fins and skin;	*Aeromonas liquefaciens*	Add anti-bacterial drugs to feed
Edwardsiellosis	*Choman-byo*		Intestines swollen	Bacterial infection	Add anti-bacterial drugs to feed
Gill disease (Columnaris)	*Era-byo*	Fig. 11.5	Gills raw and rotten	Bacterium *Chondrococcus columnaris*	Add anti-bacterial drugs to feed
Branchionephritis	*erajin-en*		Gills swollen, raw, rotten etc. and kidney inflammation	Unknown	Add salt to pond (5–7‰ in salinity)
White spot disease (Ichthyophthiriasis)	*Hakuten-byo*	Fig. 11.6	Tiny white spots on outside of body	*Ichthyophthirius multifiliis*	Pump salt water into pond
Myxidium disease	*Myxidium-byo*		Round white expanse of cysts with spores on skin	Protozoan [*Myxidium*]	Unknown (remove dead eels)
Crippled body disease (Plistophorasis)	*Beko-byo*	Fig. 11.7	Body kinked or misshapen	Protozoan *Plistophora anguillarum*	Unknown (remove dead eels)
Gas disease	*Kiho-byo*	Fig. 11.8	Bubble on head (elver) or fin (adult)	Too much O_2 or N_2	Add new water to pond or remove eels to better condition pond
Anchor worm disease	*Ikarimushi-byo*	Fig. 11.9	Parasitic crustacean attached inside mouth of eel	Parasitic copepod *Lernaea cyprinacea*	*Masoten* and bleaching powder kill larvae by sprinkling on pond; kill eggs and larvae by pumping salt water into pond

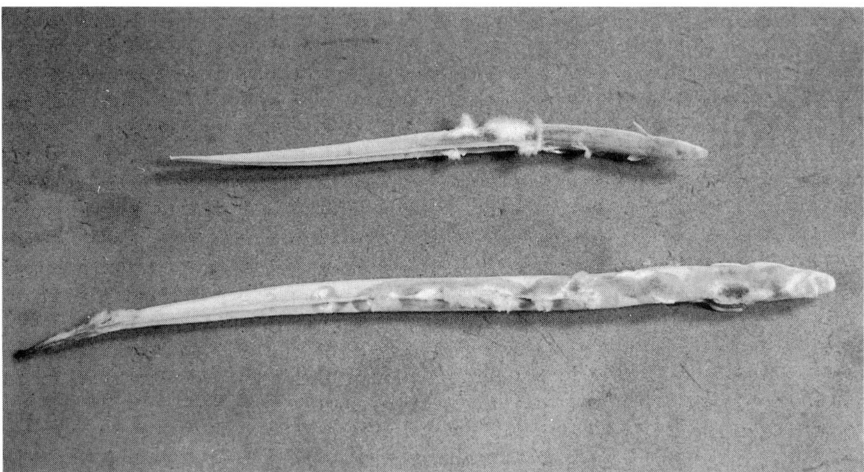

Fig. 11.1 Fungus disease (*Watamakuri-byo*). Several hundred t of eels were killed by this disease in Japan during 1966 and 1968.

Fig. 11.2 Eels killed by fungus disease floating at the edge of a pond.

Fig. 11.3 Sea gulls picking diseased eels from a pond.

struck the main eel pond area of Japan (around Hamanako to Yaizu district) and several hundred tons of eels died.

The disease breaks out during periods when the water temperature is 15–20°C and thus in Japan two peaks of infection occur, one in spring and one in autumn. Patches of white fungal growth develop and spread on the bodies of eels and cause death after one or two weeks. Fifty to seventy per cent of eels in a pond may die. However, in hot water culture using a greenhouse, which is now established in almost all Japanese farms, the disease does not occur.

To prevent outbreaks of white fungus disease, elver ponds used to be sprinkled with some nitrofuran compounds or prophylactic anti-bacterial drugs were added to the feed. However, the use of nitrofuran compounds has been prohibited in Japan because they were suspected of being carcinogenic.

To cure saprolegniasis the following have been used, but most of the diseased eels cannot be cured:

- for one week add anti-bacterial drugs to the feed, e.g. *Daimeton* at 150 g/t of body weight of fish;
- put Malachite Green into the pond water, four times at 10-day intervals (0.2 ppm), or put Methylene Blue into pond water for three days (2 ppm).

However, in Japan Malachite Green and Methylene Blue are now restricted to situations where no alternative treatment is available or for sterilizing elvers not intended for consumption.

AEROMONIASIS OR RED DISEASE (*HIREAKA-BYO*)

This bacterial disease mainly affects large eels and occurs when water temperatures are 18°C or higher in the summer. The disease (Fig. 11.4) causes the fins and body to become raw and red, and internally the intestines, liver and kidneys are affected. The disease sometimes originates from the use of decaying food, so care must be taken to keep the diet fresh.

To cure infected eels, anti-bacterial drugs are added to the feed for 5 to 7 successive days (100–200 g of medicine per 1 t of eels).

EDWARDSIELLOSIS OR SWOLLEN INTESTINE DISEASE (*CHOMAN-BYO*)

This is a bacterial disease in which the kidneys and liver become swollen and the abdomen distends. Several anti-bacterial drugs are effective.

Fig. 11.4 Red disease (*Hireaka-byo*).

GILL DISEASE (*ERA-BYO*)

This is caused by bacteria which attack the gills and cause them to rot and death follows (Fig 11.5). The addition of anti-bacterial drugs to the feed is said to cure gill rot in some cases.

BRANCHIONEPHRITIS (*ERAJIN-EN*)

This causes gills to be swollen, raw, rotten etc. and kidneys to be inflamed. The cause of this disease is unknown at present. The cure for this disease is also not settled but it is good to keep the pond water to 5–7‰ in salinity. During 1969 and 1970, this disease struck the main eel pond area (about 90%) of Japan, killing 2600 t of eels.

WHITE SPOT DISEASE (*HAKUTEN-BYO*)

This is caused by *Ichthyophthirius multifiliis* (Fig. 11.6). The cure for white spot disease is to pump salt water into the eel pond and let the eels live in salt water for a few days.

Fig. 11.5 Gill disease (*Era-byo*).

Fig. 11.6 White spot disease (*Hakuten-byo*). The white spots are colonies of the protozoan parasite *Ichthyophthirius multifiliis*.

MYXIDIUM DISEASE (*MYXIDIUM-BYO*)

Spores of a parasitic protozoan, *Myxidium*, settle on the skin, develop into a round white expanse and grow larger. The course to take for this disease is to remove the diseased eels and burn them.

CRIPPLED BODY DISEASE (*BEKO-BYO*)

The protozoan parasite *Plistophora* attacks the muscular system, causing the victim to become crippled and misshapen (Fig. 11.7). In small eels, the diseased part of the muscles can be seen through the skin as a milky colour.

GAS DISEASE (*KIHO-BYO*)

If oxygen or nitrogen in the water becomes excessive, bubbles appear on the eel's head and occasionally on the fins (Fig. 11.8). The bore hole water, which brings an excess of N_2 in the pond, should be checked prior to use. The cure for this disease is to run new water in the pond or to work the water agitator for reducing O_2.

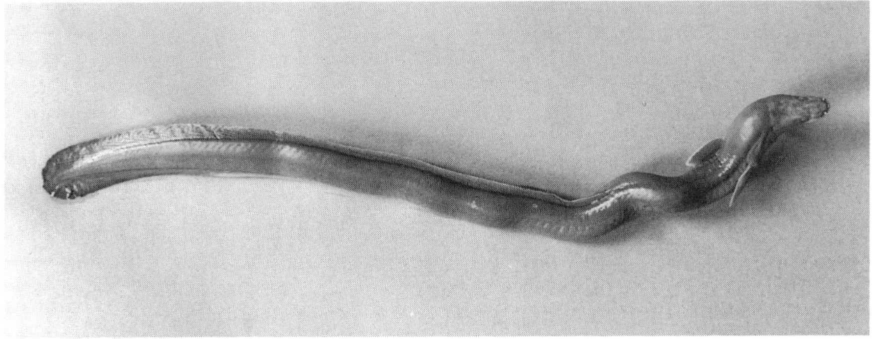

Fig. 11.7 Crippled body disease (*Beko-byo*) is caused by protozoa attacking the muscles.

Fig. 11.8 Gas disease (*Kiho-byo*) occurs in elvers.

ANCHOR WORM DISEASE (*IKARIMUSHI-BYO*)

This important condition is caused by the parasitic copepod *Lernaea cyprinacea*. This animal spends its life attached to the inside of the mouth of the eel, where it greatly hinders the eel's ability to feed, and attached to other parts of the outside of eels' bodies (Fig. 11.9). The heads of the parasites are embedded in the flesh of the eel, sucking out the juices. Heavy infections cause death by preventing the eels from feeding.

Adult *Lernaea* release up to 5000 eggs each, which hatch into free-swimming larvae, which swim about until they meet an eel, when they hang on and the story is repeated. During one summer the life

Table 11.2 Drugs approved for use on eels in Japan.

Drug	Trade name	Source	Indication	Application and dose	Withdrawal period (days)
Oxytetracycline HCL	*Terramycin* powder for fish* (10% powder)	Pfizer[1]	Edwardsiellosis	50 mg/kg of fish weight per day in feed	30
Oxolinic acid	*Parazan* D for fish (5% solution)	Tanabe[2]	Edwardsiellosis	6 hours dip at 5 ppm	25
	Parazan for fish* (5% powder)	Tanabe	Red fin	5–20 mg/kg of fish weight per day in feed for 4–6 days	20
			Red spot	1–5 mg/kg of fish weight per day in feed for 3–5 days	
			Edwardsiellosis	20 mg/kg of fish weight per day in feed for 5 days	
Piromidic acid	*Panos* for fish (pure powder)	Dainippon[3]	Edwardsiellosis	10–20 mg/kg of fish weight per day in feed for 5–7 days	20

Miloxacin	*Oskacin* for fish (5% powder)	Sumitomo[4]	Edwardsiellosis	5–30 mg/kg of fish weight per day in feed for 3–7 days	20
Sulfamonomethoxine and its Na salt	*Daimeton* powder for fish (10% powder) *Daimeton* natrium for fish (pure powder)	Daiichi[5]	Red fin	150–200 mg/kg of fish weight per day in feed	30
Trichlorfon	*Masoten* solution 20% (20% solution)	Bayer Japan[6]	Anchor worm	Add to pond water at 0.2–0.5 ppm	5
	Masoten powder for fish (80% powder)			Add to pond water at 0.2 ppm	5

* Many products based on the same generic drugs are available from other sources in Japan.
(1) Pfizer Pharmaceuticals Co., Ltd., P.O. Box 226, 1–1, Nishi-shinjuku 2 chome, Shinjuku-ku, Tokyo 163, Japan.
(2) Tanabe Seiyaku Co., Ltd., 2–10, Dosho-machi 3 chome, Chuo-ku, Osaka-shi 541, Japan.
(3) Dainippon Pharmaceutical Co., Ltd., 6–8, Dosho-machi 2 chome, Chuo-ku, Osaka-shi 541, Japan.
(4) Sumitomo Pharmaceuticals Co., Ltd., 2–8, Dosho-machi 2 chome, Chuo-ku, Osaka-shi 541, Japan.
(5) Daiichi Pharmaceuticals Co., Ltd., Matsuda Yaesu-Dori Bldg., 10–7, Hachobori 1 chome, Chuo-ku, Tokyo 103, Japan.
(6) Bayer Japan Ltd., 10–8, Takanawa 4 chome, Minato-ku, Tokyo 108, Japan.
These drugs are licensed for use on eels in Japan: different drugs are available in other countries, and further information should be obtained in each country from the ministry dealing with medicine registration. In Britain, the appropriate authority is the Veterinary Medicines Directorate, Woodham Lane, New Haw, Weybridge, Surrey KT15 3NB.

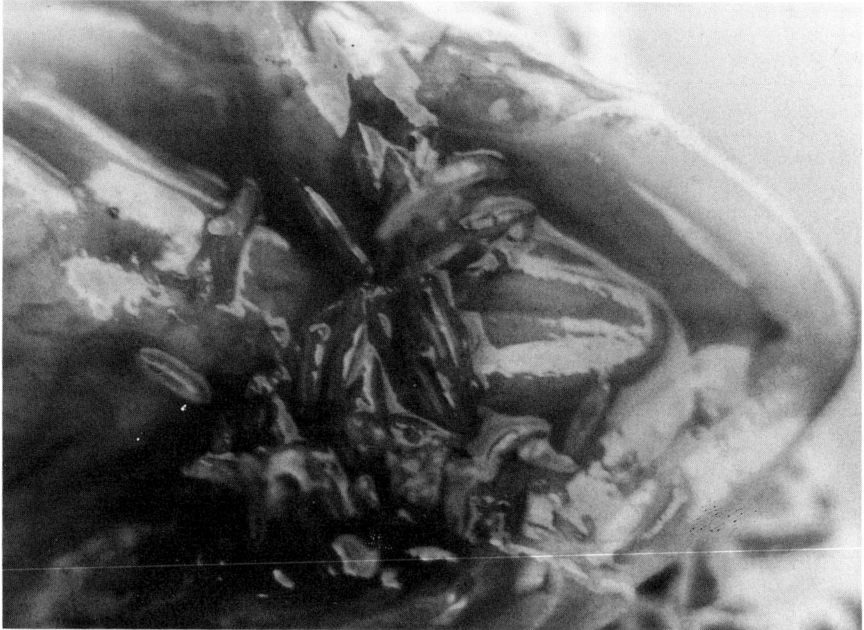

Fig. 11.9 Anchor worm disease (*Ikarimushi-byo*). Many parasitic anchor worms (*Lernaea cyprinacea*) attach themselves inside the mouth of the eel where they suck the blood and prevent the eel from feeding.

cycle is repeated about five times. The most suitable propagation condition is with a water temperature of 14–32°C; below 14°C the parasites do not reproduce. Since *Lernaea* also live on carp etc., care should be taken that any carp kept in eel ponds are disinfected before being put in.

To cure anchor worm disease, sea water should be pumped into the ponds and the eels should be allowed to live in sea water for 3–4 days. This kills eggs and larvae of the parasite. Calcium chloride powder sprinkled over the surface of infected ponds just after dawn kills eggs and larvae floating near the surface but disperses and does not harm eels at the bottom.

Masoten added to ponds at concentration of 0.2–0.5 ppm kills larvae of *Lernaea*, which float on the water surface at dawn owing to their phototaxis. For prophylaxis, *Masoten* can be applied routinely twice a month during the summer months.

OTHER PESTS AND DISEASES

In addition to the above, various internal and external parasitic nematodes and trematodes also attack eels, but are not important.

In Europe, live eels stored in perforated barges in fresh water at Maldon, Essex, in summer get a disease called 'red sickness' in which the fins become tinged red, the body hard and stiff as if rheumatic and death follows. To prevent this disease the barges are towed into sea water and moored there for two hours in every fortnight during the summer when temperatures are above 18°C.

Medicines customarily used in Japan are listed in Table 11.2. The use of Dipterex was prohibited by law in Japan from 1971, because it is an agricultural preparation; *Masoten* for fish, with the same active ingredient (Trichlorfon) is used.

12 Marketing eels

Eels have to be caught many times in their lives in order to separate fast and slow growers and move them to larger ponds as they grow, as well as the final harvesting prior to sale. At all stages it is important that the eels should not be damaged. There are three main ways of catching eels in ponds.

(1) By draining the pond and catching the eels in a 3 m long net bag tied over the outlet pipe (Fig. 12.1). In case of a bad-drainage pond, the vertical pump in the resting corner (for drainage) is used; the net with bamboo curtain is set in front of the resting corner in order to harvest eels. (Fig. 12.2)
(2) By scooping a net around eels gathered at the pond feeding place (Fig. 12.3).
(3) By drawing a seine net across the pond (Figs 12.4, 12.5, 12.6).

All these methods give good results and at any given time that which is most appropriate, can be used. To avoid damaging the eels, scoop and drag nets should be hauled very slowly.

Eels are endangered by sudden changes of temperature of 4°C or more, thus if eels are being transferred from one pond to another, the

Fig. 12.1 Bag net tied over the exit pipe of a pond to harvest the eels by draining the pond.

Fig. 12.2 Eels can be caught by setting this net in front of the exit sluice. The bamboo curtain acts like the wing of a seine to guide the eels into the net.

temperature of the receiving pond must be checked in advance and, if necessary, adjusted (by means of changing the water flow) to be similar to that of the donor pond.

Concentrating eels in a net causes the oxygen content of the water to become much reduced in that area of the pond and the dense mass of eels to gasp for oxygen. To prevent mortality, netting and subsequent size grading of the live eels must be done under conditions of maximum oxygen. Thus it is done near the main water inlet in the shallowest part of the pond (in the 'resting pool' if there is one), and splasher aerator machines are positioned nearby and run at full speed. If the net is hauled in another part of the pond for some reason, the eels are quickly transferred to a big square keep net and are moved to the best oxygenated part of the pond as soon as possible. (Figs 12.7 to 12.10)

Captured eels can be sorted into different sizes using mesh containers that allow eels of a certain thickness to escape (this is used chiefly in grading fingerling size, 20–30 cm eels, see Table 12.1, Fig. 12.11), or

Fig. 12.3 Using a scoop net to catch eels. No food is given during the previous two or three days and now as the eels come hungrily to the food the men scoop them up.

Fig. 12.4 A seine can be used to harvest eels.

Fig. 12.5 The seine net drawn tight. Care must always be taken to handle the eels carefully and avoid injuring the skin.

Fig. 12.6 Netted eels are put into plastic baskets to be lifted on to the sorting table.

Fig. 12.7 The keep net is placed in a corner of the pond close to the water inlet where the oxygen is at its maximum.

Fig. 12.8 Eels being transferred from the seine net to the keep net.

Fig. 12.9 A scene at the keep net.

Fig. 12.10 Fingerlings in keep nets await sorting while a splasher-aerator machine keeps up the oxygen supply.

Table 12.1 Mesh size of wire mesh containers for grading fingerlings, and the size of eels which they permit to escape

Mesh size	Size of eel that can escape the mesh
7 mm	7 g
9 mm	13 g
11 mm	19 g
13 mm	37 g

by tipping larger eels onto a sloping wooden sorting table whence workmen flick different sized eels into different containers (Figs 12.12, 12.13, 12.14).

It takes eels about three days to digest food they have eaten. Thus when eels are harvested for marketing they must be kept without food for about three days before they are packed for sending to market. This process is called *ikeshime* in Japanese. If this process is not carried out, eels will continue to defaecate after packing, their containers will become polluted and the eels will arrive in poor condition, if not dead.

To complete the three-day starvation gut-cleaning period the eels are stored in containers under conditions of plentiful oxygen supply. There are several methods:

Fig. 12.11 Fingerling size eels are graded into different sizes by using mesh containers.

- Eels are placed in polyethylene perforated tubs under showers in a special shed. The baskets sit in piles on top of each other and the water trickles down from one to the other (Fig. 12.15). A similar method is used in Billingsgate Market, London.
- Eels are placed in polyethylene baskets called *Doman* (Fig. 12.16), about 20 kg eels in each basket, and the baskets are placed in a concrete sluiceway full of fast-flowing fresh water (Fig. 12.17), or in the mid-pond (Fig. 12.18).

After three days the eels will have lost about five per cent of their weight owing to the complete passing out of their intestinal contents. Starvation beyond this period will cause additional continuing loss of weight, especially if the temperature is warm, which causes the eels to be active (Fig. 12.19). Stored silver-stage eels lose weight much more slowly than do brown-stage eels and at 10°C can be stored two or three months with little weight loss.

Eels are sent to market by one of four methods:

(1) Alive, journeys of more than 8 hours in cool temperatures, in simple boxes packed with a little ice.

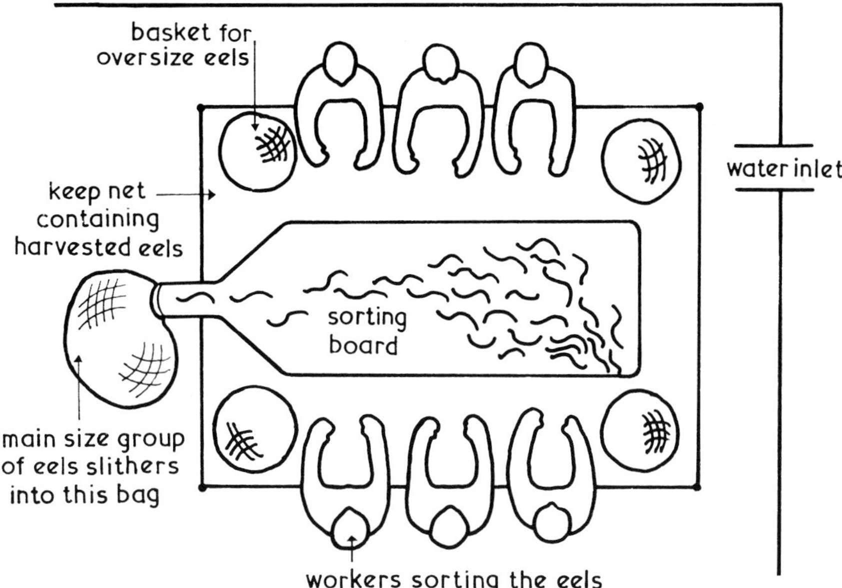

Fig. 12.12 An arrangement for sorting eels to different sizes. Eels to be harvested are held in a keep net staked in the resting corner of the pond. Eels are then scooped from the net onto the sorting table. Big eels are flicked into special baskets and undersized eels are flicked back into the keep net to be grown further. The majority of the eels slither down the board into a bag net tied over the end.

Fig. 12.13 The harvested eels are sorted and the small ones returned to the pond for further growth.

(2) Alive, in double polyethylene bags with oxygen and ice, dispatched by train or truck.
(3) Alive, in aerated tanker lorries for journeys of up to one week.
(4) Dead, quick frozen and glazed.

For short journeys in cool climates, such as when sending some eels from Somerset three hours by train to London, eels can simply be put in wooden boxes lined with wet sacking and with a little ice on top. They will arrive in good condition. However, if there is some unexpected delay, they may die.

In Japan, live eels are sent from the culture ponds to Tokyo and Osaka by truck in double polyethylene bags packed with oxygen and ice (Figs 12.20, 12.21). Packing procedure is as follows:

(1) Take two strong polyethylene bags, put one inside the other.
(2) Place 10 kg of live eels in the inner bag.
(3) Place some cubes of ice in the bag to lower the metabolic activity of the eels.
(4) Inflate the bag with oxygen from a cylinder and tie it.
(5) Place bag of eels in a cardboard carton. (Fig. 12.21)

Packed in this way, eels will easily survive for 30 hours, and in cool weather or for shorter journeys more eels and less ice can be placed in each bag.

Fig. 12.14 Another type of sorting table.

Tanker lorries are used to carry live eels between different parts of Europe. These tanker lorries are huge (Fig. 12.22). These have been developed chiefly by the eel firm Joh. Kuijten of Spaarndam, Netherlands. The most modern lorries weigh 15 t empty and carry 15 t of live eels swimming in 15 t of water. A compressor supplies air bubbles continuously into each tank and the water more or less serves only to keep the eels wet.

Each 4 days the tanker crew drain the water from the tanks and fill up with new water. In this way they can go for up to 14 days. To bring a load of live eels from Greece to Hamburg is easy. Lough Neagh's eel catch is taken to the continent in 20 hours using such tankers. They even collect eels from the USA; the live eels are loaded into the tanks, the lorry drives to a port, is lifted on the deck of a cargo ship, sits on the deck with compressor running during the voyage, is lifted ashore in Europe and drives to its destination. For a Tanker storage barge see Fig. 12.23.

Fig. 12.15 Sometimes the starvation cleansing of the eels is done by stacking the perforated plastic tubs under showers of trickling water.

Fig. 12.16 The *Doman* storage baskets are made of polyethylene.

Fig. 12.17 Harvested eels are kept 3 days in running water without food before dispatch to market. This process, called *Ikeshime*, allows the eels to complete digestion of food in their stomachs and empty all excretions from their guts. They are held in *Doman* baskets in front of the water inlet of the pond.

Fig. 12.18 *Doman* baskets of eels left for the starvation cleansing in mid-pond.

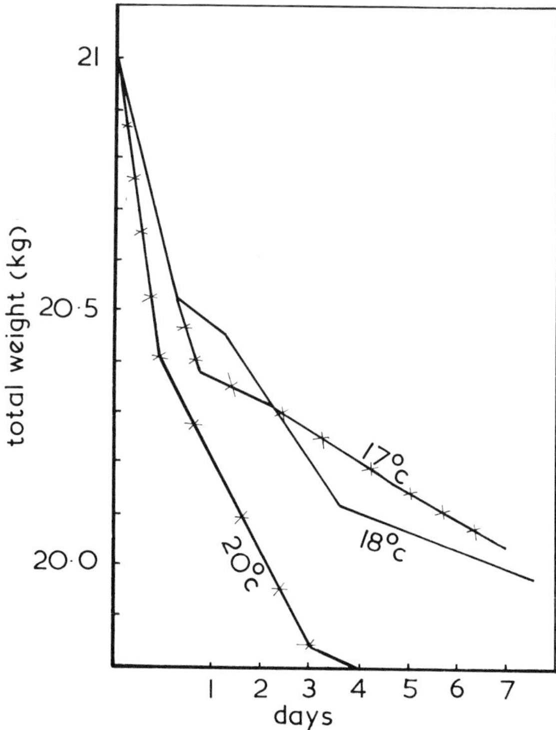

Fig. 12.19 A weight loss of about five per cent occurs during starvation cleansing. This graph shows the rate of weight loss during starvation cleansing at temperatures of 17, 18 and 20°C.

Fig. 12.20 10 kg of eels in a double polyethylene bag inflated with oxygen.

Fig. 12.21 Polyethylene bag inflated with oxygen is put in carton.

Quick-frozen eels are prepared in much the same way as other fish. They are frozen either whole or gutted, cleaned and then quick-frozen either in blocks in a plate-freezer or in an air-blast and stored at minus 20°C.

Eels have a high fat content and therefore the bodies must be protected from oxidation, which causes rancidity. This is achieved by glazing each eel or block of frozen eels and then wrapping the eel or block in polyethylene and sealing the wrapping. Stored at minus 20°C, such eels will keep in good condition for six months.

There are several ways to kill eels:

(1) Put the eels in a deep container, sprinkle on plenty of salt and leave them for two hours. The salt destroys their slime and the eels die of asphyxiation.
(2) Put the eels in a tank of fresh water and stun them with an electric shock.
(3) Put the eels in trays in a cold store overnight. The eels may not be killed but are torpid and can easily be handled for gutting etc.

Newly killed eels must be cleaned of slime by washing in cold water and scraping. A one per cent ammonia solution helps remove slime.

Fig. 12.22 Eels being discharged from a tanker able to carry 15 tons of eels. Similar lorries carry cargoes of live eels from USA to Europe, the aerators working during the voyage. Photo G. Williamson.

Fig. 12.23 A special barge with perforated sides and bottom stores 60 t of eels at Maldon, England. The barge is kept in a canal but to prevent disease is moved into sea water for two hours every two weeks. Photo G. Williamson.

To gut washed eels, the belly is slit open with a knife from the throat to one inch beyond the anus. Using the back of the knife, the guts are scraped out, taking care to remove the gall bladder without breaking it, and the gutted eel is washed carefully to remove all traces of slime and blood.

Eels can be gutted satisfactorily in quantity by machine. When hand gutting, sawdust or salt sprinkled on the eel enables the worker to get a firm grip on the fish; sometimes a rough cloth is used or the hands are dipped in dry salt for the same purpose. Weight loss during gutting may be 5–10 per cent. Heads are not removed.

Dead eels cannot be deslimed using salt, but instead are put for a few hours in a cold store, where the formation of ice on the outside of their bodies loosens the slime.

13 Month by month management of ponds in the Hamanako area of Japan

JANUARY

Pond temperatures are low, about 5–8°C, and over-wintering eels are hibernating buried in the mud. Inflow of bore hole water into ponds is cut off, save for 1–2 hours per day, in order not to disturb the hibernation of the eels by introducing water of a different temperature. No food is given. Blooms of phytoplankton (*Flagellata*) may occur during this month. Malachite Green at 0.2 ppm is added to ponds to prevent white fungus disease (but see p. 75 for a note on current restrictions in Japan).

Ponds which have been harvested and elver ponds are dried out at this season and lime is sprinkled on the bottoms to aid the sunshine to disinfect the ponds. The first run of elvers is generally caught in January. They are placed in oxolinic acid solution (*Parazan D*) of 5 ppm for six hours to kill any wild bacteria etc. on their bodies.

FEBRUARY

Water temperatures are 3–6°C. This is the coldest month of the year. To prevent marked temperature changes between day and night from disturbing the hibernating eels, water levels in ponds can be raised to make the ponds deeper than usual. Market prices for eels are good at this season. This month being the settlement term of the year, eels are often harvested by draining the pond, but it is dangerous to do so because eels are then subject to injury. The eel's bed should be improved by giving improving chemicals to the bottom soil.

MARCH

Pond temperatures are 10–12°C. The pH value should be kept at 7–8. Blooms of phytoplankton, especially *Scenedesmus*, occur as the water warms up for the spring and this causes bad water quality and the pond becomes green. By increasing the rate of water flow through the pond, the transparency of the water should be kept at 15–20 cm. White fungus attacks and other diseases may start this month but as yet there is no effective preventive method. Malachite green should be sprinkled on the pond, eels closely inspected for disease and any dead eels removed. This is the main month for catching elvers.

APRIL

Pond levels are lowered to speed warming by the sun. Water temperatures are 13–20°C and elvers are fed in their warmed greenhouse tanks. They double their weight in 20 days. In mid-month elvers are caught and sorted into large and small size groups. Adult eels quit their hibernation and are active and hungry. They become eager to feed when temperatures rise to about 18°C but the start of full-scale feeding is usually delayed until this period in order that the ponds shall have attained consistently warm temperatures. Epidemics of white fungus disease often occur this month, simultaneous with the start of feeding. Some antibiotic or sulfa drug added to the feed is believed to reduce the risk of bacteria attacks.

MAY

Water temperature is 20–22°C. Attacks of the white fungus disease die down as the water reaches temperatures in the twenties. Fingerling eels feed vigorously eating artificial feed of about 2.5 per cent, or raw fish of about ten per cent of their body weight per day. Blooms of phytoplankton (*Microcystis*) bring about good water conditions causing eels to have good appetites. If blooms of water fleas occur the water quality declines and this causes the eels to lose their appetites for the duration of the bloom. *Masoten* should be added to the pond at a concentration of 0.2 ppm to kill the water fleas and anchor worm larvae. Salt water is pumped into the pond for 3–4 days to kill anchor worm young stages. To increase the efficiency of feeding, elvers should be sorted as often as possible, though not necessarily every day, after the first selection.

JUNE

Water temperature reaches 25°C. This is the height of the rainy season in Japan and the weather is hot, humid and much rain falls. In the early mornings, oxygen levels in eel ponds fall dangerously low, causing the eels to rise gasping to the surface. The reasons are:

- dense phytoplankton is present and uses up much oxygen at night;
- decomposition of excretion and organic detritus on the pond bottoms use up much oxygen; and
- at these high temperatures of 25°C plus, water cannot hold much oxygen in solution.

To provide extra oxygen during the danger hours between 08.00 and 18.00 splasher aerator machines in the ponds are switched on and water inflow increased. Daily variations in rainfall and sunshine greatly affect the oxygen level in the ponds and hence the eels' appetites vary from day to day during the hot summer months. Some eels in their second growing season reach marketable 150 gm size this month, and can be harvested with nets.

Swollen intestine disease may occur in this month and can be treated by adding anti-bacterial drugs and vitamin E to the feed. Since oxygen levels are so low at the bottom of ponds, it is vital that no uneaten feed be left in ponds or it will lie on the bottom decomposing anaerobically and further poisoning the water. Lime and ferric oxide (trade name: *Manken*) are scattered on ponds to break down organic residues now accumulating on pond bottoms.

A typical daily routine is as follows:

Time	
06.00	Collect and analyse water samples from each pond; take general look at all ponds.
08.00–10.00	Feed adults and elvers.
08.00–15.00	Feed elvers continually.
10.00–13.00	Carry out any regular daily application of chemicals or medicines to ponds; remove dead eels; declog any filters, outlet sluices.
	Do any special tasks of an occasional nature, e.g. catching eels to sell, monthly application of chemicals to ponds; maintain equipment and pond; sample fish to check growth and health.
15.00	Skim off green algae that may have grown on surface of covered elver ponds (it can become thick in a few hours' sunlight).
17.00	Collect and analyse water samples from ponds.
19.00	Make evening check of all ponds.
20.00–06.00	On hot nights check ponds by torchlight and switch on splasher machines if oxygen level runs low and eels are seen gasping at the surface.

JULY

Water temperatures are 27–28°C and the eels feed vigorously. The hot summer is the season when the eels put on most weight. The clarity of the pond water goes through continuous cycles from transparent to turbid (visibility less than 15 cm) owing to successive blooms of phytoplankton. By adjusting the inflow of water, the water quality is maintained in healthy limits. When dense blooms occur, more water is let in and vice versa. Rotifer epidemics may occur. *Masoten* is put in the ponds to kill anchor worm larvae.

The Japanese festival called *ushi-no-hi* (day of the ox) occurs this month and everyone likes to eat eel *kabayaki*. Thus many eels 120–150 g are harvested by net to take advantage of the big demand at festival time.

AUGUST

Water temperature is maximum of the year, 30–32°C. At this temperature only a little oxygen can dissolve in the water and each night the eels have to come to the surface to gasp for breath. To compensate for the lack of oxygen, the splasher aerator machines are run all day long and maximum volumes of new oxygenated water are admitted to the ponds. Red disease and gill disease occur in the hot temperatures of this month but incidence of swollen intestine disease declines.

SEPTEMBER

Water temperatures are 20–27°C. Heavy feeding of eels continues. Plankton changes from *Microcystis* to *Oscillatoria* and *Scenedesmus*, the water in the pond being green. On some days blooms of *Anabana* occur which greatly reduce the appetites of eels. In ponds containing brackish water, the water turns brown or thick soy sauce colour because of blooms of diatoms, but when the water is settled in colour the eels reveal great appetites again. Splasher aerator machines continue to be run 24 hours a day to keep oxygen levels up. This month is typhoon season in Japan.

OCTOBER

Water temperatures are 18–20°C and the cool of autumn causes the eels' appetites to decrease and vary from day to day. Phytoplankton of a brown colour dominates the ponds. The splasher aerator machines are run each night and sometimes all day too.

NOVEMBER

Water temperatures are 12–15°C. Eels of all sizes gradually lose their appetite. Reduced amounts of feed continue to be given until finally no eels come to the feeding point because they are all in hibernation. Some anti-bacterial drugs and mixed vitamins are mixed with the last week's feed to protect the eels from disease over the winter hibernation. Water depth is increased during winter to afford more protection to the eels. Lime and ferric oxide (*Manken*) is sprinkled on the ponds to make the condition of the bottoms good during the winter.

DECEMBER

Water temperatures are 7–10°C. The eels hibernate, except a few which may visit shallow edges of the ponds on sunny days. December–March are the main economical months for harvesting eels and this is done by netting or draining ponds.

Also during this month the year is reviewed and the all-important planning for next year takes place.

14 How to catch and treat elvers

Elvers are the starting point of eel culture. Since eels breed in the sea and since no one has yet persuaded eels to breed in captivity, eel farmers must obtain stocks of elvers by catching wild elvers that enter river mouths at their accustomed time each year. The elvers of *Anguilla japonica* and of all the eel species of the temperate zones then enter rivers in huge numbers. Big rivers like the Loire in France receive something between 50 million and 100 million elvers each year.

Elvers do not arrive at all parts of the coast at the same month. For instance in Europe *A. anguilla* elvers enter rivers in Western Spain in December–January and the River Severn, Britain, in April–May (Table 14.1).

There are two reasons for this. Firstly, the areas nearest to the spawning ground receive elvers earlier than the remote regions. Secondly, elvers are not stimulated to migrate toward river mouths until river temperatures rise above about 8°C. The annual wave of *A. anguilla* elvers arrives at the edge of the European continental shelf in November–December onwards each year, about one year and ten months after being spawned, but the rivers do not become warm enough to attract the elvers until later.

Roughly speaking, elvers of all eel species colonizing temperate lands arrive at the coast some time in the spring. The size of the immigrant elvers varies from species to species. After metamorphizing from leptocephali at sea, elvers do not feed until they enter fresh water, and actually lose weight during the period while they wait at sea for rivers to warm up. Elvers entering a river late are thinner and shorter than are the early runs in the same river. To reach Italy, elvers have to travel along the Mediterranean and do not arrive until some months after others have reached Spain. Elvers enter rivers mainly on spring tides, i.e. in a series of waves of two week intervals. They swim only at night and keep close to the river bank, not in mid-stream, and enter on each night's rising tide. Usually they swim at the surface, sometimes in an almost solid procession. Jim Milne, famous elver fisherman of the River Severn, once caught 25 kg of elvers in one scoop of his net. That is about 87 500 individual elvers!! During daylight elvers remain hidden in the mud and under stones. If they come to a weir or waterfall they get around it by wriggling up the damp moss on either side of the falls. To catch elvers there are several methods (Figs 14.1 to 14.8):

- Using scoop nets at night at a river bank, to which the elvers are attracted by hanging a bright light.

Table 14.1 Times of elver (*Anguilla anguilla*) runs in parts of Europe

Place	Age of elvers entering rivers (spawning is in February)		Month of main elver run	Size of elvers (no. per 1 kg)
	Years	Months		
Spain and Portugal, Atlantic coasts	1	10	Dec.–Jan.	2 700
France, Biscay and Brittany coasts	2	0	Jan.–Mar.	2 800
British Isles, North Sea coasts, Scandinavia	2	3	Apr.–Jun.	3 500
Italy, west coast around Pisa	2	9	Nov.	4 000
Egypt, River Nile	3	0	Jan.–Mar.	4 500

Fig. 14.1 A dip net used for catching elvers in the river Severn, England. Elvers sweep in on high tide and after the tide ebbs they start moving upstream. At promontories they hug the banks and best catches are made there. Sometimes a single dip of the net will secure 10 kg of elvers. Photo G. Williamson.

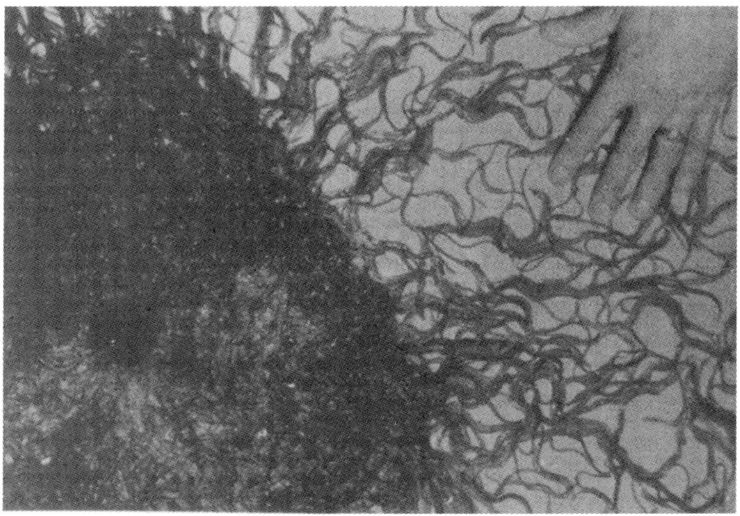

Fig. 14.2 New caught elvers. Photo G. Williamson.

Fig. 14.3 An elver catcher weighs his catch at the Epney-on-Severn collecting station. Of some 50 t caught in this river yearly about half are eaten, a quarter exported to Germany and a quarter flown to Japan. Photo G. Williamson.

Fig. 14.4 Gauze cages used by Japanese elver catchers for holding their catches. As elsewhere this involves great activity at the time but the business has expanded so much in Japan that more elvers are needed than can be supplied locally.

Fig. 14.5 Tanks for holding elvers at Epney-on-Severn. Photo G. Williamson.

Fig. 14.6 Emptying an elver holding tank. Note the square of perforated aeration hose at the bottom. Photo G. Williamson.

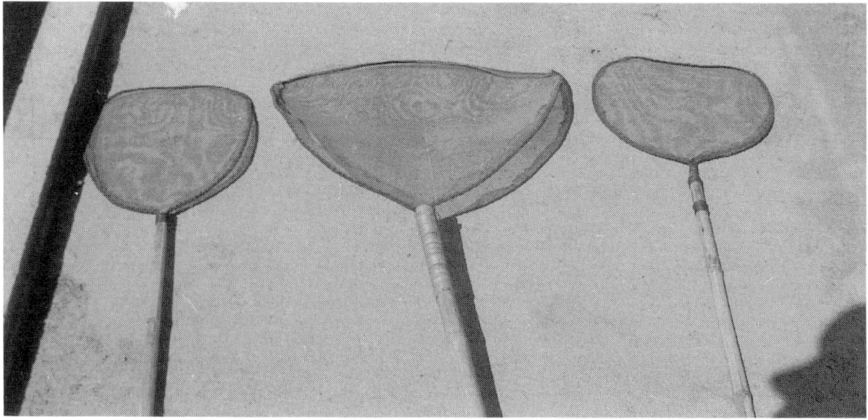

Fig. 14.7 Japanese gauze scoops for catching elvers. The net part consists of smooth surfaced metal or gauze.

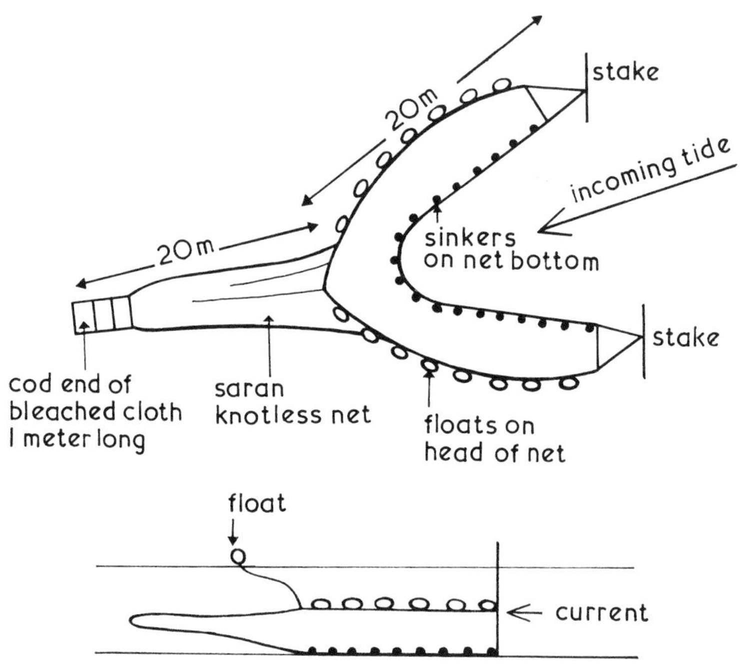

Fig. 14.8 Net scheme for catching elvers. It is staked across a river with a float attached to the headline to keep the mouth wide open.

- Setting a fine-mesh net across the width of a river and so catching all the elvers coming up on a flood tide.
- Constructing a special elver-trap at a weir or waterfall where the elvers upstream migration is obstructed.

A properly designed and built structure can catch almost every single elver that enters a river. On the River Bann in N. Ireland such a system catches an average of some 23 000 000 elvers every year.

Elvers are very delicate. Caught elvers should not be touched by hand but should be placed in a box lined with wet muslin, in the case of small numbers, or hung in mesh cages in the river (large numbers) and then taken to the eel farm within a few hours.

During December–February the coldness of the night air may injure elvers, so elvers caught at this season should be put in boxes with lids so as not to be exposed to the winter air. By March, temperatures are higher and elvers are robust and active.

At Epney-on-Severn, England, at St Nazaire on the River Loire and on the River Gironde in France there are special elver catching and storing companies. They buy live elvers from local dip-net fishermen and store the live elvers in tanks of circulating water. Most of the elvers are sent to Germany by aerated tanker trucks, to stock inland waters that receive no natural elver supply, or to Japan by air freight to stock eel culture ponds. This elver export is big business. In the River Bann, Northern Ireland, many elvers are caught but are all put into Lough Neagh.

The annual catch of elvers in these rivers is shown in Table 14.2.

Elvers can be transported by several methods:

- In special aerated tanker lorries (Fig. 14.9). About seventeen tons of brackish water are used to hold one ton of elvers.
- In gauze-bottomed boxes lined with wet muslin or in polyethylene bags with water and oxygen. (Fig. 14.4)
- For air-freight; polystyrene trays (Figs 14.10, 14.11). The polystyrene is light and insulates the elvers from sudden temperature changes. Each 0.5 kg tray holds 1 kg of elvers. The elvers are cooled to 6°C before packing to slow their activity.

The size of the Japanese eel farm industry and its demand for elvers is now so great that not enough elvers can be caught in Japan itself and agents of Japanese eel co-operatives import many tonnes of elvers by air freight from other countries. About 50 t of elvers and fingerlings

Table 14.2 Average annual catch of elvers

River	Catch
Loire (France)	200 t = 66 million elvers (many more are not caught)
Bann (N. Ireland)	78 t = 26 million elvers (virtually all are caught)
Severn (Great Britain)	50 t = 16 million elvers (many more are not caught)
Gironde (France)	50 t = 16 million elvers (many more are not caught)

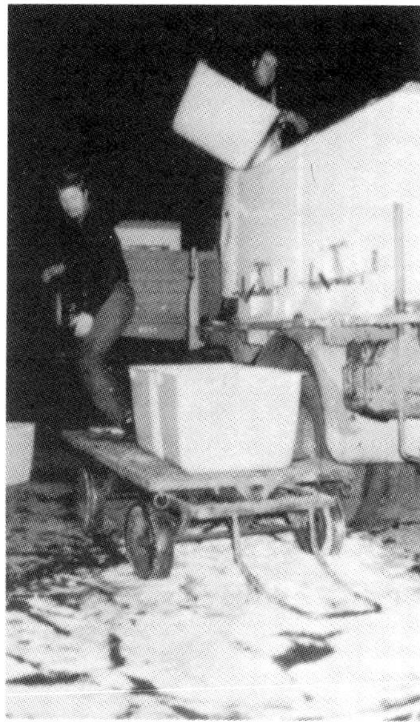

Fig. 14.9 Loading elvers from the Severn into a tanker lorry to go to Hamburg, Germany. White froth generated by the elvers covers the ground. The lorry has six tanks and an air compressor to work bubble aerators and carries 1.2 t of elvers. Salt is added to the tank to make the water half saline. The journey takes 15 hours. Photo G. Williamson.

Fig. 14.10 The latest type of trays used in air-freighting elvers from Europe to Japan. Each 0.5 kg heavy tray holds 1 kg of elvers. No ice is used; the elvers are simply cooled to 6°C before packing, put in the trays with only enough water to keep them wet and loaded on the plane. Many millions of elvers cross the North Pole like this every spring. Photo G. Williamson.

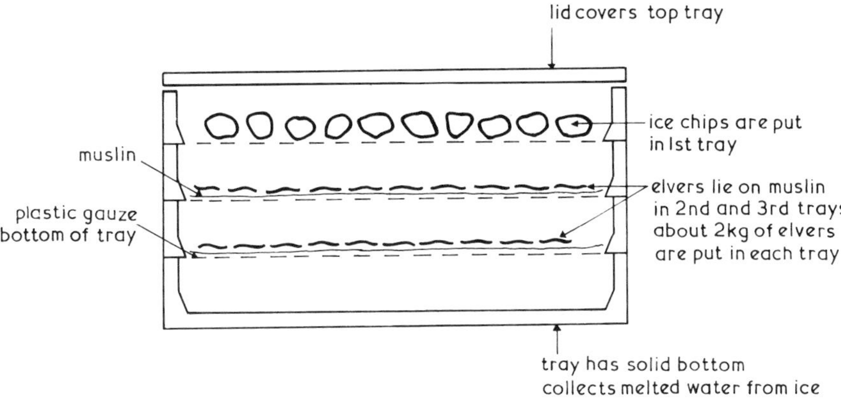

Fig. 14.11 An older type of expanded polystyrene tray for air-freighting elvers. Modern type is cheaper and simpler. (G. Williamson).

are needed each year by Japanese eel farms; Table 14.3 shows where they were obtained in 1989.

This requirement for about 50 t of elvers per year could actually be met by the elver run of a single big river like the Loire in France. Plenty of elvers arrive from the sea, including at river mouths of Japan itself; the problem is how to catch them. The only method to ensure catching all the elvers that enter a river is to build a dam right across the river mouth. Usually this is difficult and expensive, involving access for ships, politics, damage to other fisheries etc. But on one big river in N. Ireland it has been done.

A special construction for catching the entire run of elvers is used at the mouth of the River Bann, Coleraine, Northern Ireland. Something like 26 000 000 elvers enter this river every year and every one of them ends up in the two elver traps. At a point about six miles from the sea where the river is about 200 m wide a dam has been built extending right across the river. The elvers cannot ascend the fast

Table 14.3 Sources of elvers for Japanese eel farms in 1989

Country of origin	Elvers (t)		Fingerlings <15 cm (t)
Japan	40.0	*Anguilla japonica*	0.0
Taiwan	0.0	*Anguilla japonica*	1.6
South Korea	0.0	*Anguilla japonica*	0.4
North Korea	0.1	*Anguilla japonica*	0.1
China	0.9	*Anguilla japonica*	0.1
Hong Kong	6.4	*Anguilla japonica*	0.0
France	1.8	*A. anguilla*	0.0
Total	49.2		2.2

Table 14.4 Time needed to grow elvers to 200 g market size eels in natural temperature outdoors, uncovered still-water ponds in various areas of the world

Place	Approximate temperatures (°C) of still-water ponds based on mean monthly values: No. of months 23°C and above	Annual temperature range	Month-degrees annual total	Time needed for eels to grow to 200 g
Germany (Berlin)	1	0–23	150	About 4 years
England (Somerset)	0	4–22	160	About 4 years
France (Marseilles)	2	7–27	200	About 3 years
New Zealand (Auckland)	2	12–25	200	About 3 years
Australia (Melbourne)	2	8–27	200	About 3 years
Japan (Hamanako)	4	5–30	220	About 2 years
Tunisia	3	10–30	220	About 2 years
Australia (Brisbane)	6	14–31	260	About 1 year
Taiwan	6	13–31	260	About 1.5 years
Indonesia	12	25–31	330	About 1 year

flow of water falling over the central weir and are attracted to two slow-flowing trickles of water coming from the special elver chutes built at either side of the dam. A clever method is here used to trap the elvers. In the special chutes the elvers wriggle up a water-logged 'rope' of straw down which a current of water is trickling. But instead of reaching the upper river, the elvers when they come to the end of the straw drop into a trough. The secret to the working of this device is a slot in the chute which allows some water to fall onto the upper end of the straw rope, thus washing the climbing elvers off the straw into the trough beneath. Elvers always keep to the sides of a river and a loop of the straw rope on the river bed below the trap ensures that all elvers are intercepted.

Table 14.4 shows the average time needed to grow elvers to market size (200 g).

15 How to cook eels: *kabayaki*, smoked and jellied

KABAYAKI, THE JAPANESE STYLE

This is the name of the style in which eels are cooked in Japan, and the food is delicious. *Kabayaki* in Japan, smoked eels in Germany and jellied eels in London are all so tasty that it is impossible to say which is the most delicious.

Kabayaki is prepared by splitting eels, broiling lightly so as to keep the shape while streaming, then steaming the pieces and broiling them over a charcoal fire while periodically dipping them into a tasty sauce called *tare* (Figs 15.1 to 15.6). The precise composition of the *tare* used by each famous chef is secret but the basic composition is a boiled-together mixture of soy sauce (five parts), sweet *sake* (*mirin*, five parts) and a small amount of sugar.

To prepare *kabayaki* in Japan, you take a live eel, pin its head to a wooden board, then slit it up one side of the back with one deft cut.

Fig. 15.1 *Kabayaki* preparation. The eel is split open and the back-bone removed.

Fig. 15.2 Bamboo skewers are put through the pieces of eel to keep them spread flat during cooking.

Fig. 15.3 The eel, the colour of which first turns white by broiling, is then steamed and broiled a second time. Eels are taken from the steam box.

Fig. 15.4 During broiling the eel is dipped into *tare* sauce two or three times.

Fig. 15.5 The pieces are grilled both sides over a charcoal fire to produce the finished *kabayaki*.

Fig. 15.6 When traditionally served a *kabayaki* meal is laid out on a low table on *totami* mats.

Now the body of the eel is opened flat, the gut removed and the backbone is sliced off. The body is cut into pieces about 12 cm long and several bamboo skewers or metal needles are threaded through the flesh of each piece. These supports keep the meat flat during the subsequent cooking, preventing it from curling up.

The first piece is now lifted on its skewers, and boiling water is poured, first on the skin side, then on the flesh side. Over steam, the piece is then cooked lightly.

Now the piece is dipped in the delicious brown *tare* sauce and grilled over a charcoal fire first one side and then the other. The process of dipping and grilling is repeated about three times, by which time the piece of eel should be an appetising light brown colour like a kipper). It should also be completely cooked and finally should be giving off mouth-watering aromas.

And there you have *kabayaki*; eat it in true Japanese style with rice.

Formerly the only places where first grade *kabayaki* eel was obtainable in Japan were certain famous restaurants specialising in it. Now, however, *kabayaki* is also cooked in automatic cookers and can be bought by housewives in supermarkets in vacuum-sealed packs. A number of attractive and savoury recipes have been evolved in the western world for cooking eels and are available in cookery books. Here are some simple standard methods for general treatment.

SMOKED EELS

Eels are usually hot-smoked after de-sliming and gutting as described in Chapter 12. The smoking method varies considerably from country to country and the method given here was described by Horne and Birnie (1969). The eels are killed and cleaned, or properly frozen eels are thawed, and then immersed in brine for ten minutes using 270 g salt to one litre of water.

The brined eels are threaded on 0.5 cm diameter rods or spears by pushing the pointed end of the rod through the throat from side to side. The rod may be of stainless steel or wood dowelling. Small lengths of stick are placed between the belly flaps to keep them apart; this allows smoke to penetrate the belly cavity.

The eels are dried, smoked and cooked during the smoking process. The filled rods are hung in a kiln and the eels are smoked for 1 hour at 35°C (95°F), ½ hour at 50°C (120°F) and finally for one hour at 73°C (170°F). This gradual increase in temperature permits reasonably uniform drying throughout the thickness of the fish; when the temperature is raised too quickly the eels may become case-hardened, particularly when the fat content is low; that is, the skin becomes dry and hard, but the flesh remains wet. The eels should lose about 15–20 per cent by weight during the smoking operation.

The eels are removed from the kiln and allowed to cool before packing; otherwise moulds may form. They are brushed lightly with edible oil if necessary and wrapped in transparent plastic film before packing in boxes.

The finished produce is cooked and ready to eat. The flesh should have a good smoky flavour with only a slight taste of salt; the texture should be firm and buttery, but not too tough. The shelf life of the finished produce is about 3–4 days at chill temperature. 100 lb of fresh, whole eels yields about 60 lb of hot-smoked eels.

SMOKED EELS FOR CANNING

The eels are killed and cleaned as described earlier and immersed in 80 per cent brine for 15 minutes. The eels are threaded on rods hung in the kiln in the same manner as before. They are smoked for one hour at 35°C (95°F) one hour at 50°C (120°F) and finally for one hour at 73°C (170°F), the eels are removed from the kiln, allowed to cool and cut into pieces the length of the can.

The pieces are packed into cans, which are filled up with vegetable oil heated to (230°F) and then sealed and heat processed at 102°C (230°F). A 200 g (7 oz) oval can takes about 1 hour. With this method there is very little shrinkage of the meats in the can. For Australian activity see Fig. 15.7.

Fig. 15.7 Australian smoked eels. The eels are de-slimed, gutted, dipped in brine and then smoked.

JELLIED EELS

The manufacture of jellied eels is a traditional process about which there is little authentic information, and the following hints are given as a guide.

The eels are killed, gutted and cleaned as described earlier, and cut into pieces about 1½–2 inches (3.8–5.1 cm) long.

The pieces are next cooked by adding them to boiling water, bringing the water back to the boil, adding about 1 lb of salt for every 25 lb of eel, and then simmering until the flesh is soft enough to be pushed off the bone with the fingers. Cooking time will vary depending on the size of the eels, the season of the year and the area of capture. Some experience is required to decide when the eels are properly cooked, but about 10 minutes actual boiling is typical for yellow eels. Silver eels have a thicker, tougher skin and require somewhat longer, and they should be overcooked rather than undercooked to make the skin soft enough to eat. Next, cold water is added to bring the oil to the surface and the oil is skimmed off. The cooked pieces and the hot liquor are next poured into large bowls containing gelatine dissolved in a small amount of hot water, usually about a 10 per cent solution, sometimes with a small amount of added vinegar. The amount of gelatine solution needed depends on the condition of the eels and their natural capacity to gel; again experience is necessary to get the recipe right. Once the mixture has cooled, the pieces in jelly are packed into cartons for fresh consumption. Waxed cardboard, plastics and aluminium foil have all been used successfully as containers. Shelf life can be up to 2 weeks at chill temperature; waxed carton packs may have a shorter shelf life since there is sometimes some reaction between the gelatine and the wax coating.

Another recipe for jellied eels recommends the use of a salt-vinegar solution and a rather longer cooking time. Water containing two per cent vinegar and three per cent salt, together with 2 oz of spices to the gallon (12.5 g/litre), is brought to the boil and 2-inch (5.1 cm) pieces of skinned eel are added. After that addition the mixture is brought back to the boil and then left to simmer for about 45 minutes. The pieces are then put in large bowls to cool and a weak gelatine solution is added if there is insufficient natural jelly.

16 Recent developments in eel culture in Japan

All the main eel culture areas in Japan use still water culture. The reason is simple. Large volumes of pure water are not available in rivers and, instead, water supplies are obtained from underground bore holes. This bore hole water cannot be used for running water ponds; there is not enough of it and it is cold, 15–20°C.

In still water ponds the sun gets the opportunity to heat up the water greatly, even up to 32°C in the height of summer, and thus eels can grow well. By artificially heating the elver ponds starting in March, the number of months each year in which the eels live in water 23°C and above is increased from 3 to 7 months.

The Japanese have tried two further methods of eel culture. Using hot water from the plentiful geothermal hot springs, attempts have been made to warm ponds using this free heat. Unfortunately hot springs do not occur at any of the established eel culture localities and this good idea has not caught on. Culture of eels in sea water has also been tried. The eels grow well, but their taste is reported to be inferior to eels grown in fresh water.

The main recent developments occurring in eel culture in Japan are:

- Increasing use of central heating pipes to warm fingerling ponds during spring time. This increases the length of the annual growing season.
- Construction of more and more ponds. Modern ponds are only about 1000 m^2 in size, much smaller than those built in former years.
- The density of eels in ponds is being continually increased. More and more delicate control of water quality accompanies this practice.

During 1969 and 1970 branchionephritis (*Erajin-en*) (Chapter 11), struck the main eel pond area in Shizuoka Prefecture. As a consequence, the production came down to half of the previous level. The disease germinates over winter and spring, when the water temperature is <23°C. During this period, if the water temperature could be raised artificially to 27–30°C, the propagation of the disease could be inhibited. It still accounts for half of all disease-related eel losses in Japan.

With the help of greenhouse cultivation, if elvers are released into the pond by the end of the year, market size eels can be produced by the following summer. Table 16.1 shows the progress of eel production in recent years using greenhouse cultivation. After 1975, most of the

Table 16.1 Japanese eel production using the greenhouse method

Years	New elver release in pond (t)	Market size eel production (t)	Progressive ratio of eel production
1975–1979	366.4	143 519	391.7
1980–1984	401.8	179 763	447.4
1985–1988	196.0	152 610	778.6

eel culturists had complete facilities but were not at first familiar with their proper use. Over the years the utilization techniques have been improved and have become more sophisticated (Figs 16.1 to 16.4). Hence the ratio of market size eel production to elver release has become nearly 800:1. This development in eel culture has resulted in economies in fuel consumption and use of electric power. While eels are imported from Taiwan at lower prices, nevertheless the culturists can achieve a more stable supply.

BUILDINGS

The size of a greenhouse is decided by the size of the eel-rearing pond and determines the type of roof structure used. Usually humidity

Fig. 16.1 Inside a greenhouse; the water in the pond is circulated by paddle splasher machines.

Fig. 16.2 A filtration unit for pond water.

Fig. 16.3 A dried out pond. A drain pipe can be seen in the centre and a paddle-splasher machine at the left edge.

Fig. 16.4 A group of greenhouses.

inside the building is very high because the nursery pond is filled with warm water. Farming and roofing materials should therefore be rust-resistant.

The roofs are divided into three types according to shapes, barrel roof, pitched roof and dome-type roof, all of which are designed to let as much sunshine as possible into the building.

The barrel type roof is a simple framework: the roofing frame of galvanized iron is covered in vinyl sheeting. A typical simple barrel roof may be used for an eel-rearing building, but this structure is not recommended because even in a short span there are many pillars inside the building, which hinder the staff in their work. For an efficient eel-raising facility a long span of up to 12 or 13 m may be possible using a steel truss or H-shape beam frame to support the barrel roof.

A pitched roof consists of a framework of galvanized iron, covered in vinyl sheets, and again supported by a steel truss or H-shaped frame. This type of roof system is typically used for frame structures with a long span of up to 25–30 m, which can accommodate ponds of any shape or size.

A dome-type roof also uses a galvanized iron framework covered with vinyl sheeting. This type of roof is sometimes use for large buildings such as gymnasiums and enables the culturists to build the eel-raising facility as large as possible.

The buildings also require a suitable exterior wall to maintain the internal thermal conditions required. Sometimes vinyl sheet is used, with clips securing it in place (Fig. 16.5).

HEATING SYSTEMS

There are currently a number of sources for supplying heating in Japan. Most use petroleum oil, coke, coal, natural gas or industrial waste, or involve water heated by heat pumps, geothermal energy (i.e. hot springs), and thermal effluent. The heavy oil boiler is the most important method.

Either hot water boilers or steam boilers are used.

Hot water boilers supply, hot water which is pumped through a heating pipe installed in the bottom of a pond (Fig. 16.6). This water then circulates and returns to the boiler, creating a closed system whereby the pond water is heated indirectly by the boiler (Fig. 16.7).

Steam boiler systems are of two kinds. In the first, the steam generated in the boiler is discharged directly into the pond (Fig. 16.8). In the second the water in the pond is heated through a heating pipe, with condensed water being returned to the boiler for subsequent use. The steam boiler method requires a smaller diameter

Fig. 16.5 A greenhouse with a pitched roof and walls of double vinyl sheets.

Fig. 16.6 Heating pipes installed on the bottom of a pond, using hot water from boiler.

Fig. 16.7 A hot water boiler.

Fig. 16.8 Pipes from a steam boiler which heat the water by means of direct discharge.

of piping for the pond compared with the hot water boiler. Furthermore heat-proofing is simpler and heat efficiency is high, around 90 per cent.

Although the cost of such a facility is comparatively high, generally around 2–3 times that of the hot water boiler), it is very suitable for use in large scale ponds.

To determine the output required of a boiler, trial calculations for the energy loss from the pond should be made without considering the energy input from solar heat. The heat loss from the surface of the boiler house and from the bottom of the pond and that due to drainage should be determined. The pond is generally drained so as to get rid of excess eel feed and excreta. This facilitates some exchange of pond water to maintain a constant pond volume.

The heat loss in the Shizuoka area, per hour, for a 330 m^2 pond during a severe winter (air temperature is 3.0 to 5.0°C), was calculated as 159 180 kcal. In this case the height of the boiler house is 3 m, the quantity of water supplied and drained is 3 t per hour, the temperature of the water supplied is 15.0°C and the ground temperature is 5.7 to 8.0°C. The energy required during a severe winter will be about 160 000 kcal per hour; however, about 100 000 kcal will suffice if the house is enclosed in a double vinyl cover. In addition, the heat loss due to people coming and going and the ventilation due to draughts may be estimated to be about 4000 to 5000 kcal per hour. Furthermore, the heat required to raise the water temperature has to be taken into consideration. In order to raise the water temperature by 1.0°C, in a pond of 330 m^2, 60 cm deep (i.e. 200 t of water), will require 200 000 kcal. Therefore, a boiler with a capacity of 100 000 kcal will be required in order to raise the temperature of water by 1.0°C in two hours.

This calculation was made strictly with the premise of existing severe winter conditions. Hence, an allowance must be made for an increase between 20 and 30 per cent when determining the size of boiler.

17 Developments in eel culture outside Japan

TAIWAN, SOUTH KOREA AND CHINA

These countries followed Japanese eel culture methods and production is increasing. The annual production is about 52 000 t in Taiwan, 3000 t in South Korea and 10 000 t (including 3000 t wild catch) in China (see Table 1.1).

EUROPE

Among the countries of Europe many activities concerning eels are in progress. Europe eats 25 000 t of eels annually and the demand still increases. The main developments are:

- Experimental and commercial attempts to culture eels.
- Research into new techniques of fish culture, especially the use of continuously circulated and purified water in heated insulated tanks.
- Expansion of air-freight export of elvers to Japan. A record quantity of about 75 tons was sent to Japan in 1972 from France, England and Italy though the quantity varies and in 1989 was only 1.8 t (Table 14.3).
- Continued research into many aspects of eel biology including a prospective voyage by the Danish research vessel *Dana* to the Sargasso to attempt to catch the first-ever spawning adult eels and thus complete the research begun by Johannes Schmidt in 1904.

The main problem of eel culture in the eel eating countries of Northern Europe, Germany, Netherlands, Denmark and England, is that they are all too cold. Temperatures of 23°C and over, at which eels grow fast, seldom or never occur (Table 14.4). In open air ponds eels take 4 years to reach market size. Japanese eel ponds are in the latitude of Gibraltar and Tunisia.

There are several possible ways to tackle the problem as shown below, and all are being investigated. Which ways will be commercially successful is another question.

(1) Use polyethylene multi-span covers to cover still water ponds, trap solar heat and prevent heat loss due to evaporation and convection (Fig. 17.1).

Fig. 17.1 Multi-span polyethylene greenhouses can be used in temperate countries to warm the water. Illustrated are some designed at the Lea Valley Experimental Horticulture station in England.

(2) Build still water ponds with insulated bottom and sides (panels of styrofoam), covered above with greenhouse sheeting and artificially warm the water.
(3) Build running water ponds, insulated and heated as above, in which the warm water is not discharged but is purified (physically by filters, chemically by resins) and recirculated, thus eliminating heat loss. Such ponds will be small volume hexagonal or circular structures. Alternatively these ponds could be fed with warm water effluent from a power station without recirculation.
(4) Abandon attempts to culture eels in the cold north and instead develop open air still water culture in warmer areas such as around the Mediterranean, perhaps even in the Caribbean.

AMERICA

The USA in the one country in the world where wild eel catches could be greatly increased. Eels occur from Maine to the Gulf of Mexico and are abundant in many areas. Possibly 10 000 t a year could be harvested but at present only about 400 t are caught.

Canada has plenty of eels in the St Lawrence which are fully harvested. Owing to the cool temperatures, growth rates are slower than in much of Europe.

Eel culture has many academic followers in America but it will probably not develop commercially at present since few Americans like to eat eels.

NEW ZEALAND

A sizeable fishery for wild eels has recently developed in New Zealand and 1000 t were caught in 1987. Formerly only Maoris caught and ate the two species of eel in New Zealand and even fought tribal wars over the fishing rights of various areas. Now the eels are caught by many small companies. The eels are nearly all exported frozen to Germany, the Netherlands, Britain and Japan and earn foreign exchange.

Two species of eel occur in New Zealand, the long-finned eel, *Anguilla dieffenbachi*, which is peculiar to New Zealand, and the short-finned eel, *A. australis*, which also occurs along the eastern coast of Australia. Both species probably spawn near New Caledonia, from where a warm current sweeps down to the Australian coast and round to New Zealand. The elvers reach New Zealand in the spring, September and October each year.

Anguilla dieffenbachi grows to a huge size. Silver females average 6 kg weight and may reach 20 kg. They may stay 30 years in fresh water before migrating to sea, far longer than the European eel (Table 2.2). The fat content of silver stage specimens is about 25 per cent.

Anguilla australis is a modest-sized species that looks very similar to the Japanese and European eels. Female silver eels migrate after about 22 years in fresh water, and have a weight of about 600 gm and a fat content of 16 per cent (Table 2.2).

Near Christchurch on the South Island is one of the phenomena of the eel world, a brackish lagoon called Lake Ellesmere, about 260 km^2 (100 square miles) in size and 2.3 m deep, which is full of eels. An average of 300 t of eels, containing both species, is caught annually.

Lake Ellesmere does not have a permanent connection with the sea but is separated from it by a bank of gravel through which water percolates. Elvers can evidently wriggle through the gravel into the lake but silver eels cannot get out. Men catch the silver eels by digging ditches in the gravel from the lake towards the sea during the silver migration season. The silver eels detect the increased flow of water out of the lake and at night swim in huge numbers into the ditches in an attempt to get to the sea; but instead they are caught. Similar lagoons exist in Newfoundland, Canada.

To encourage the eel industry, efforts are being made by the Fishing Industry Board of New Zealand (general manager Mr R.W. Dobson) and Victoria University, Wellington (Dr P.H.J. Castle, Zoology Department). The main objectives are to expand the catch of wild eels to the maximum sustainable yield, but not beyond; to encourage development of eel culture; and to discover more about the biology of New Zealand eels.

Eel culture will probably get the best results using the still water method and will be located in the northern parts of North Island around Hamilton, Auckland and the Northland, where water temperatures are warmest; temperatures of 23°C and above occur during two months each year (Table 14.4).

Gratitude is due Mr J.S. Campbell and Mr P. Chapman of New Zealand Fishing Industry Board and Dr P.H.J. Castle, Mr P.R. Todd and Mr D. J. Jellyman of Victoria University Zoology Department for information and data supplied.

AUSTRALIA

Two species of eel occur along the coasts of Eastern Australia and commercial quantities are found in Victoria, Tasmania and New South Wales. About 250 t per year are caught, mainly by fyke nets, and there is interest in the possibility of starting eel culture. A State licence to start an eel farm in Tasmania was granted early in 1973.

Anguilla australis is the most important species and occurs from western Victoria north up to Brisbane. It is important because it is the most abundant species (about 200 t caught per year) and also because in size and plain colour it closely resembles the European and Japanese eel. The greatest biomass lives in Victoria and nearly all Australia's catch is taken here at present. The eels are exported frozen to Europe.

Anguilla reinhardti is a big mottled species and occurs from Tasmania right up to northern Cape York and also on New Caledonia. It grows to a large size, is at present used only for shark bait, and only about 10 t are caught annually.

Both species probably spawn to the east of New Caledonia. Elvers enter rivers in Queensland in winter (June–September), in New South Wales (July–October) and Victoria in spring (August–November). These records refer to unidentified elvers, since as yet no research biologist has carried out the painstaking task of separating in each sample the elvers of the two species.

Anguilla australis reach silver stage at about 600 g, the same as in New Zealand, and *A. reinhardti* silver eels are very large, about 5000 g. Silver stage sizes are given in Table 2.2, maximum sizes are given in Table 2.1.

It seems probable that the wild population of eels in Australia is not high and that wild eel harvests cannot be increased dramatically. The climate in Queensland is fully as warm as in Japan and would be suitable for eel culture (Table 14.4).

Thanks are due to Mr P. C. Pownall, editor of *Australian Fisheries* and Dr D. Buckminster of Victoria Freshwater Fisheries Department, for data supplied.

TROPICAL COUNTRIES

It is in tropical countries that the majority of *Anguilla* species live (Fig. 2.5). Indonesia is the ancestral home of the genus and no fewer than nine species live there to this day. Other species live in East Africa, Madagascar, India, Burma, New Guinea and the Pacific Islands.

It is not believed that large catches of wild eels can be taken in these tropical areas because, although many individual eels occurs, some very large, the total biomass is small.

Culture of eels is also unlikely to develop, for the simple reason that the people in tropical countries need all the protein they can get. They cannot afford the luxury of raising carnivorous animals which destroy protein. Producing 1 kg of eel flesh requires that the eels eat about 7 kg of the flesh of some cheaper fish.

What is required in tropical countries is biological research into the lives of the tropical eel species. The biologist who does this has all the discoveries listed in Chapter 3 open to him and will become a famous man. As yet hardly anything is known about the lives of the tropical eel species.

18 Check list of requirements for starting an eel farm

CHARACTER OF THE PERSON IN CHARGE

(1) Have you a sound business sense and knowledge of financial procedures? Do you understand the scientific details of eel biology?
(2) Are you of a quiet temperament and manually skilful at practical engineering such as making cement structures? If you are a high-powered tycoon type you will quickly get bored with eel farming and quit.

CAPITAL FINANCING

(1) Have you sufficient capital to cover the capital costs of starting a farm?
(2) Have you in addition sufficient capital to pay the running costs of the farm?

PRELIMINARY ENQUIRIES

(1) Have you read all the available literature on eel culture? You should make visits to commercial fish farms and government culture research stations to see what they look like and meet people who can advise you. Can you afford a two-week trip to Japan itself to see the real eel culture industry?
(2) Have you sought the advice of experts in the following fields; running a business, banking, accountancy, real estate, engineering, biology?
(3) To set up an eel farm you will need permission from your local river board and county planning authority. These people have much helpful advice to give: the earlier you contact them the more they can help you.
(4) Make out a detailed costing of your operation. Estimate costs high and production low.

PHYSICAL FEATURES OF SITE

(1) Water supply volume. Is at least 40 million gallons (18 000 m^3) available in the driest months of the year? It can come from a river, lake, as well as borehole.

(2) Water supply quality.
   (a) Is the water supply free from possibility of dangerous pollution? Check the whole upstream part of any river you propose to use to see if there are any factories and sewage outlets which may cause trouble. Ask farmers what insecticides they use and when. Water need not be crystal clear; murky water is fine. If wild eels occur in the water, that is a proof it is of good quality.
   (b) How can you detect pollutants entering your farm?
   (c) How can you prevent pollutants entering your farm?
   (d) What recourse have you in case of loss due to pollution?
   (e) Can technical assistance be obtained for detection of pollution and at what cost?
(3) Flooding: Are you sure that the site can never be flooded?
(4) Area: You need about 4 acres (1½ ha) to build a 40 t per year production farm and 10 acres (4 hectares) more for future expansion. (See Table 8.1) Can you obtain this land by purchase or by lease?
(5) Services: Road access and electricity are vital. A telephone is handy.

BIOLOGICAL FEATURES

(1) Elvers: Where will you get your elvers? At what cost? In which months are they available?
(2) Diseases: If eels become diseased, what facilities have you for diagnosing which disease is responsible? Where is the nearest parasitologist/fish disease expert who can help you? What costs may be involved?
(3) Feeding: What is the present price per tonne of artificial feed? What is the price of raw fish and how much per tonne does it cost to freeze and store raw fish?

MARKETING

(1) Your profit largely depends on the price at which you sell your eels. You need marketing know-how and skill. Will you sell your eels in bulk at a low price to a firm that will collect them? Or will you supply smaller amounts direct to retailers at a higher price? Which firms have you contacted?
(2) Do you want to smoke your eels yourself and pack them?

These and many other questions must be considered before undertaking the venture but, if properly based, there is ample evidence to indicate that the steadily rising price of eels because of ever expanding demand offers a growing attraction for profitable investment and enterprise.

19 References and bibliography

A peculiar situation exists regarding literature on fish culture owing to the fact that most persons actively operating farms are in the business for profit and do not wish to reveal their trade secrets. Therefore books tend to be written by biologists and academic people and to be vague, academic or out of date. Useful scientific and commercial information is none the less to be found in print, and more will be published as the years go by.

A huge number of articles on aspects of aquaculture have been published in Japanese and dozens more are published each year. They are published in Japanese because they are intended for local use and are mostly produced by the prefectural experimental stations. Several good textbooks exist, and recently several practical layman's guides to commercial culture of certain species have been published; books on yellowtail and eel culture are available so far. In the following list, the English translation of each title is given first.

Publications of particular practical use to eel farmers are marked with an asterisk.

BOOKS AND ARTICLES

* Bertin, L. (1956) *Eels: a Biological Study.* Cleaver-Home Press, London.

Bruun, A.F. (1937) Contributions to the life history of the deep sea eels: Synaphobranchidae. *Reports of the* Dana *Expedition*, **9**, 1–31.

Ege, V. (1939) A revision of the genus *Anguilla* Shaw. *Reports of the* Dana *Expedition*, **16**, 256 pp.

Egusa, S. (Ed.) (1990) *Consideration on Fish Diseases* (In Japanese: *Gyobyo ronko*). Kosei-sha Kosei-kaku, Tokyo.

* Evans, A. (1973) *Aquariums.* Foyle's Handbooks, Palmetto, Pubns, London.

Horne, J. & Birnie, K. (1969) Catching, handling and processing eels. *Torry Advisory Note*, No. 37. MAFF, London.

Huet, M. & Timmermans, J.A. (1972) *Textbook of Fish Culture: Breeding and Cultivation of Fish.* Fishing News Books, Oxford.

* Iizuka, S. (1971) *Eel* (In Japanese: *Unagi*). Nosan Gyoson Bunka Kyokai, Tokyo. (A practical guide to eel culture)

* Iizuka, S. (1979) *Latest Eel Culture* (In Japanese: *Unagi no saishin*

yoshoku-ho). Nosan Gyoson Bunka Kyokai, Tokyo.
Inaba, D. (1966) *Study of Freshwater Propagation* (In Japanese: *Tansui zoshoku gaku*). Kosei-sha Kosei-kaku, Tokyo.
* Inaba, D. (Ed.) (1976) *Fisheries Propagation* (In Japanese: *Suisan zoshoku*). Kosei-sha Kosei-kaku, Tokyo.
* Inaba, S., Matsui, I., Tsunogai, H., Aoe, H. & Ogami, H. (1971) *Eel* (In Japanese: *Unagi*). Midori Shobo, Tokyo.
* *Iversen*, E.S. (1976) *Farming the Edge of the Sea* (Second Edition). Fishing News Books, Oxford. (A survey of coastal culture)
Jespersen, P. (1942) Indo-pacific Leptocephalids of the genus Anguilla: systematic and biological studies. *Reports of the* Dana *Expedition*, **22**.
* Kafuku, T. & Ikenoue, H. (1983) *Modern Method of Aquaculture in Japan*. Elsevier Science Publishers, Amsterdam, and Kodansha, Tokyo.
Kawamoto, N. (Ed.) (1965) *Elements of Aquaculture Methods* (In Japanese: *Yogyo gaku so ron*). Kosei-sha Kosei-kaku, Tokyo.
Kawamoto, N. (Ed.) (1967) *Aquaculture Methods for All Species* (In Japanese: *Yogyo gaku kaku ron*). Kosei-sha Kosei-kaku, Tokyo.
* Matsui, I. (1970) *Theory and Practice of Eel Culture* (In Japanese: *Yoman no riron to jissai*). Revised edition. Nippon Suisan Shigen Hogo Kyokai, Tokyo. English translation published in 1980 by Balkema, Rotterdam.
* Matsui, I. (1971) *A Book about Eels* (In Japanese: *Unagi no hon*). Marunouchi Shuppan, Tokyo. (Contains essays, legends and miscellaneous writings about the eel)
* Matsui, I. (1972) *Eel Study* (In Japanese: *Man gaku*). Volume 1, Biological Study; Volume 2, Technical Study on Culture, Kosei-sha Kosei-kaku, Tokyo.
* Milne, P.H. (1972) *Fish and Shellfish Farming in Coastal Waters*. Fishing News Books.
* Nemoto, K. (1988) *Facilities of Eel Culture* (In Japanese: *Shisetsu yoman gijutsu*). Kosei-sha Kosei-kaku, Tokyo.
* Nomura, M. (Ed.) (1982) *Aquaculture Methods* (In Japanese: *Tansui yoshoku gijutsu*). Kosei-sha Kosei-kaku, Tokyo.
* Oshima, Y. & Inaba, D. (Eds) (1971) *Eel* (In Japanese: *Unagi*). Lecture on Fish Culture, Volume 7. Midori Shobo, Tokyo.
Schmidt, J. (1932) *Danish Eel Investigations during 25 years, 1905–1930*. Carlsberg Foundation, Copenhagen.
* Williamson, G.R. (1971) Lessons in fish farming from Japan. *Fishing News International*, May and June 1971.

MAGAZINES AND JOURNALS

American Fish Farmer. P.O. Box 1900, Little Rock, Arkansas 72203, USA.

The Canadian Fish Culturist. Department of Fisheries of Canada,

Ottawa, Canada.

* *Electrical Surplus.* R.F. Winder Ltd, Belgrave Electrical Works, Stanningley, Leeds, UK. (For advertisements of second-hand pumps, etc.)

* *Farmer's Weekly.* 161 Fleet Street, London, EC4. (For contacting landowners and for advertisements of construction products, etc.)

FAO Fish Culture Bulletin. Fisheries Division, Biology Branch, Food and Agriculture Organization of the United Nations, Via delle Terme di Caracalla, Rome, Italy.

* *Fish Culture* (In Japanese: *Yoshoku*). Midori Shobo, Tokyo. (Monthly trade magazine)

Fish Culture Times (In Japanese. *Yogyo taimusu*) Yogyo Taimusu, 1–2–6, Iwamoto-cho, Chiyoda-ku, Tokyo. (A twice-monthly trade paper)

The Fish Farmer. 1378 Livermore Avenue, Livermore, California 94550, USA.

* *Fish Farming International.* Heighway Ltd, 35–39 Bowling Green Lane, London, EC1R 0DA. (For current developments in fish culture the world over)

* *Grower.* Grower Publications Ltd, 49 Doughty Street, London, WC1.

* *The Progressive Fish Culturist.* Superintendent of Documents, US Government Printing Office, Washington DC 20402, USA.

Appendix 1 Status and organization of aquaculture in Japan

Aquaculture is an important and growing industry in Japan. In 1988, the annual total production of fisheries was 12 785 000 t comprising 1 327 000 t of shallow sea culture, and 99 000 t of inland water culture, one thousand million yen in value, indicated according to species of fish. Of this eel culture amounted to 39 558 t, valued at 584.3 million yen.

Fry supply, research, legislation and subsidies all contribute to the aquaculture industry and many firms make equipment specially designed for aquaculture.

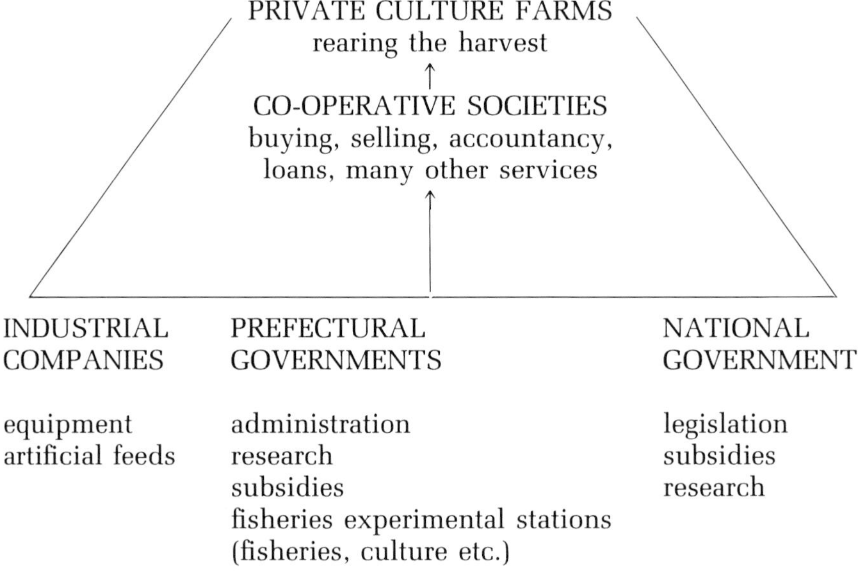

At the present time, and for several decades past, a feature of Japan has been the very heavy expenditure of public funds on fisheries. Part of this money is used to support aquaculture, although the majority is devoted to normal fisheries. There are at the present time in Japan 86 main and 62 branch fisheries experimental stations (including breeding centres and culture experimental stations); 23 fisheries universities and colleges; 2030 staff members (including biologists, research chemists and technicians); and 484 advisers engaged in guidance to fishermen and fish farmers in Prefectural Government employ.

Undoubtedly this massive Government backing has helped the industry greatly in the past. Now, however, the industry seems to make progress chiefly by its own efforts. The best aquaculture men are farmers themselves, the staffs of breeding centres and culture experimental stations, and excellent biologists at experimental stations and university fisheries faculties.

PRODUCTION

The marketed harvest is mostly produced by many small companies or individual farmers. These are usually members of co-operative societies, which serve many functions; marketing the produce of members, purchasing fry, in bulk for members, providing accountancy services, acting as the farmers' mouthpiece when dealing with government etc. Only a few large companies engage in fish farming.

FRY SUPPLY

Fry of rainbow trout, carp etc., seed of abalone, scallop, etc. and young prawns are produced artificially at 37 special breeding centres, since they cannot be caught wild in adequate numbers. To operate a breeding centre is a sophisticated and expensive business, so these centres are operated by the Prefectural Governments and have scientific staffs. From the centres the fry are sold at subsidised prices to co-operatives or individual farmers. Young invertebrates and fry such as prawns, blue crabs and rock fish etc. are raised in National Government Farming Centres and are subsequently released directly into the Inland Sea to increase wild stocks of these species.

The breeding centres are of key importance and as the list of cultured species expands, their work will increase. They also serve as demonstration farms where private fish farmers and others can see the latest techniques and equipment in use.

RESEARCH

About 148 prefectural experimental stations exist scattered all over Japan, sometimes more than three in a prefecture. At 135 of these, aquaculture is among the subjects studied. The aim of these stations is practical and local, and many are doing almost identical work. These stations are in contact with the co-operative societies of fish farmers in their areas and can arrange visits to individual farms. Culture research of a more academic and usually less practical nature is carried out at the two fisheries universities and 21 university fisheries faculties. The *Suisancho* (Fisheries Agency) operates eight fisheries research laboratories, six of them doing culture research.

Free advice for farmers is available at any of the 135 prefectural fisheries experimental stations and nine fisheries research laboratories of *Suisancho* and 23 fisheries universities and this is an important aid. Research reports in Japanese on aquaculture are produced regularly by most of these stations, laboratories and universities.

A huge amount of information on aquaculture has been published, but largely in Japanese.

ADMINISTRATION

The *Suisancho* (Fisheries Agency, Ministry of Agriculture, Forestry and Fisheries) in Tokyo is the body which administers aquaculture at a national level, and consists of Fishery Administration Department, Oceanic Fishery Department, Fishing Port Department and Research and Development Department to provide administrative guidance, promotion of subsidies and research study. This Research and Development Department has a staff of 412 researchers, including 208 engaged in culture research.

Appendix 2 Suppliers of feeds and equipment in Japan

To service the needs of fish and eel farmers, specialities are provided by a number of reliable firms, as follows:

ARTIFICIAL FEEDS FOR MANY FISH SPECIES

Nippon Formula Feed Mg. Co., Ltd., No. 2, Yasuda Building, 3–32–13, Tsuruya-cho, Kanagawa-ku, Yokohama City.
Chubu Shiryo Co., Ltd., 14–2, Kitahama-cho, Chita-city, Aichi Prefecture.
Nisshin Flour Milling Co., 19–12, Koami-cho, Nihonbashi, Chuo-ku, Tokyo.
Nihon Nosan Kogyo K.K., Sotetsu KS Building, 1–11–20, Kitasaiwai-cho, Nishi-ku, Yokohama City.
Yeaster Co., Ltd., 726, Fukuda, Honda-cho, Tatsuno-city, Hyogo Prefecture.
Taiyo Shiryo Co., Ltd., 3–3, Shinminato, Chiba Prefecture.
C. Ito Food Mills Co., Ltd., 2–9–6, Kaji-cho, Chiyoda-ku, Tokyo.
Hayashikane Sangyo Co., Ltd., 2–10–3, Higashiyamato-cho, Shimonoseki-city.
Fuji Flour Milling Co., Ltd., 3–1–18, Seikai, Shimizu City.

FEED OILS TO ADD TO ARTIFICIAL FEEDS

Riken Vitamin Oil Co., 3–8–10, Nishi-kand, Chiyoda-ku, Tokyo.
Nippon Chemical Feed Co., Ltd., 3–6, Asano-cho, Hakodate City, Hokkaido.
Nippon Vitamin Oil Industrial Co., Ltd., 155, Nakamaruko, Nakahara-ku, Kawasaki City, Kanagawa Prefecture.
Nippon Suisan Kaisha Ltd., 6–2, Otemachi, Chuo-ku, Tokyo.

NETS, FISHING GEAR, FISHERIES EQUIPMENT etc.

Nichimo Co., Ltd., 6–2, Otemachi, Chuo-ku, Tokyo.
Taito Seiko Co., Ltd., 1–2, Higashi-shinbashi 1-chome, Minato-ku, Tokyo.
Hakodate Seimo Sengu Co., Ltd. (Tokyo Branch), Nitta Building, 8–2–1, Ginza, Chuo-ku, Tokyo.

FLEXIBLE TARPAULIN TANKS AS CULTURE PONDS

Toray Industrial Inc., 2–2–1, Muromachi, Nihonbashi, Chuo-ku, Tokyo.
Asahi Chemical Industrial Co., Ltd., 1–12, Yurakucho, Chiyoda-ku, Tokyo.
Unitika Ltd., 3–4–4, Muromachi 3-chome, Nihonbashi, Chuo-ku, Tokyo.

STC POLYCARBONATE TANKS FOR REARING LARVAE AND ROTIFERS

Teijin Chemicals Ltd., 6–21, Nishi-shinbashi 1-chome, Minato-ku, Tokyo.
Mitsubishi Gas Chemical Co., Inc., 2–5–2 Marunouchi, Chiyoda-ku, Tokyo.

STYROFOAM BARREL-SIZE FLOATS FOR NET CAGES

Sekisui Chemical Co., Ltd., 36 Mori Building, 3–4–7, Toranomon, Minato-ku, Tokyo.
Bridgestone Corporation, 1–10–1 Kyobashi, Chuo-ku, Tokyo.

VERTICAL PUMPS AND PADDLE-SPLASHERS FOR POND-CURRENT/AERATION

Yasuda Denki Kogyosho K. K., 4, Nishikicho, Toyohashi-city, Aichi Prefecture.
Hamamatsu Denko Sha Co., Ltd., 1958, Tennomachi, Hamamatsu-city, Shizuoka Prefecture.
Otsuka Tekko Co., Ltd., 1650, Kando, Yoshida-cho, Haibara-gun, Shizuoka Prefecture.

COMPLETE KITS FOR MAKING GREENHOUSES OF POLYETHYLENE SHEETING STRETCHED OVER FRAMES OF STEEL TUBES

Mitsui & Co., Ltd., 1–2–1, Otemachi, Chiyoda-ku, Tokyo.
Tsukiboshi Kogyo Co., Ltd., 11–5, 4-chome Hatchobori, Chuo-ku, Tokyo.

Appendix 3 Meanings of some Japanese words

unagi	eel
shirasu unagi	elver
byoki	disease
doman	round container for holding eels
kenkyusho	research laboratory
kumiai	co-operative society
ikeshime	method of starving eels to clean their guts
sakana	fish
shikenjo	experimental station
suisan	fishing industry
yoshoku	fish culture
yoman	eel culture
yoshokujo	fish farm
sanso	oxygen
jinko jiryo	artificial feed
-cho	sub-section of a town
-ken	prefecture
-ku	area of a city (ward)
Suisancho	Fisheries Agency, Ministry of Agriculture, Forestry and Fisheries, Japanese Government

Index

Pages, Tables and *Figures* are listed in that order

Books published by
Fishing News Books

Free catalogue available on request from Fishing News Books, Blackwell Scientific Publications Ltd, Osney Mead, Oxford OX2 OEL, England

Abalone farming
Abalone of the world
Advances in fish science and technology
Aquaculture in Taiwan
Aquaculture: principles and practices
Aquaculture training manual
Aquatic weed control
Atlantic salmon: its future
Better angling with simple science
British freshwater fishes
Business management in fisheries and aquaculture
Cage aquaculture
Calculations for fishing gear designs
Carp farming
Catch effort sampling strategies
Commercial fishing methods
Control of fish quality
Crab and lobster fishing
Crustacean farming
The crayfish
Culture of bivalve molluscs
Design of small fishing vessels
Developments in electric fishing
Developments in fisheries research in Scotland
Echo sounding and sonar for fishing
The economics of salmon aquaculture
The edible crab and its fishery in British waters
Eel culture
Engineering, economics and fisheries management
European inland water fish: a multilingual catalogue
FAO catalogue of fishing gear designs
FAO catalogue of small scale fishing gear
Fibre ropes for fishing gear
Fish and shellfish farming in coastal waters
Fish catching methods of the world
Fisheries oceanography and ecology
Fisheries of Australia
Fisheries sonar
Fisherman's workbook
Fishermen's handbook
Fishery development experiences
Fishing and stock fluctuations
Fishing boats and their equipment
Fishing boats of the world 1
Fishing boats of the world 2
Fishing boats of the world 3
The fishing cadet's handbook
Fishing ports and markets
Fishing with electricity
Fishing with light
Freezing and irradiation of fish
Freshwater fisheries management
Glossary of UK fishing gear terms
Handbook of trout and salmon diseases
A history of marine fish culture in Europe and North America
How to make and set nets
Inland aquaculture development handbook
Intensive fish farming
Introduction to fishery by-products
The law of aquaculture: the law relating to the farming of fish and shellfish in Great Britain
The lemon sole
A living from lobsters
The mackerel
Making and managing a trout lake
Managerial effectiveness in fisheries and aquaculture
Marine fisheries ecosystem
Marine pollution and sea life
Marketing: a practical guide for fish farmers
Marketing in fisheries and aquaculture
Mending of fishing nets
Modern deep sea trawling gear
More Scottish fishing craft
Multilingual dictionary of fish and fish products
Navigation primer for fishermen
Net work exercises
Netting materials for fishing gear
Ocean forum
Pair trawling and pair seining
Pelagic and semi-pelagic trawling gear
Pelagic fish: the resource and its exploitation
Penaeid shrimps — their biology and management
Planning of aquaculture development
Refrigeration of fishing vessels
Salmon and trout farming in Norway
Salmon farming handbook
Scallop and queen fisheries in the British Isles
Scallop farming
Seafood science and technology
Seine fishing
Squid jigging from small boats
Stability and trim of fishing vessels and other small ships
Study of the sea
Textbook of fish culture
Training fishermen at sea
Trends in fish utilization
Trout farming handbook
Trout farming manual
Tuna fishing with pole and line